WAS BEETHOVEN A
BIRDWATCHER?

A Quirky Look at Birds in History and Culture

DAVID TURNER

summersdale

WAS BEETHOVEN A BIRDWATCHER?

Copyright © David Turner, 2011

Illustrations by Joe Beale at Pickled Ink.

Summersdale Publishers Ltd
46 West Street
Chichester
West Sussex
PO19 1RP
UK

www.summersdale.com

Printed and bound in Great Britain by CPI Mackays, Chatham ME5 8TD

ISBN: 978-1-84953-145-0

Substantial discounts on bulk quantities of Summersdale books are available to corporations, professional associations and other organisations. For details contact Summersdale Publishers by telephone: +44 (0) 1243 771107, fax: +44 (0) 1243 786300 or email: nicky@summersdale.com.

CONTENTS

INTRODUCTION

I don't want to snipe, but the sight of politicians auspiciously hawking new policies by flying kites for them on the evening news makes me want to ululate – and I'm not larking about. They're a bunch of overpaid woodcocks, as greedy as gannets – and they should be at home by that time of night if they want to avoid being cuckolded, as my friend Mavis (more of whom later) points out.

I'm not talking cockney rhyming slang – although I will be later, to tell you how East Enders pressed the humble sparrow into service to enrich the local argot. I'm simply using the language of birds, which has been pressed into service for centuries to enrich Britain's vocabulary.

It's a great pity that the average well-educated person knows a good deal less about the country's birds than he or she did a hundred years ago, when the bulk of the population lived nearer to the land. Although Britain was already heavily industrialised by that stage, its towns and cities had not swelled into today's sprawls, which have left urban dwellers so isolated from many of the sights and sounds of nature. Birds have left their traces in words and expressions, but often we are unaware of the origins of this vocabulary. Their song has been eulogised by the great poets, but because we no longer know the voices of individual birds, it is very difficult to appreciate the writing they inspired. To forget about birds is to take one more step towards alienation from our own language and culture.

This book tries to redress this quite recent dearth of bird knowledge by taking a look at some of the most interesting birds in Britain and elsewhere in the world. But it has another purpose too: to reveal the surprisingly important role that the bird kingdom has played in the world of humans – in history and in culture.

I've done this by choosing seventy-six birds and devoting an essay of two or three pages to each, In many cases the bird in the title illustrates a broader point about birds or humanity, so I might start by talking about one bird and go on to talk about another.

I've divided the birds into eight categories, most of which are self-explanatory, but I should explain the last three. Many birds can sing but 'songsters' have a particularly pleasant song which is worth listening to – and I have made myself the judge and jury on whether a bird merits inclusion in this section (with the aid of a little guidance from a Mr Wordsworth and a Mr Shelley, acting in an advisory capacity and with no claim on my royalties). All the songsters are in the group known as the passerines, or 'perching birds', an 'order', or super-family, of birds, that accounts for a little over half the known species in the world. They include most small land birds, like sparrows and finches. 'Other perching birds' that don't have outstanding voices but have a different reason to be interesting are in a category of their own, in this book. 'Maverick birds' are species that defy categorisation. One example is the hoatzin of South America, which isn't closely related to any other birds at all. Another is the corncrake, which can be heard in Scotland and Ireland. It is a type of rail, but unlike most rails shows no interest in getting its feet wet, so I've put it in this category because it's a bold bird that dares to flout rail convention.

So read on if you want to know which bird started a war, which won a war, which became a gay icon, which was kept as a pet by Mozart, and which lent its song to a symphony by Beethoven. For centuries birds have provoked diplomatic incidents and invoked flights of genius by poets and musicians. Now it's time to have a bit of fun seeing the extraordinary, and often surprising, role which they've played for centuries in the affairs of humanity – from ancient creation myths and cave paintings to the fateful spring day in 2010 when a bird upstaged the two most powerful men in Britain.

Note: For ease of reading, one-word bird names e.g. robin, swift, bee-eater, megapode, are not capitalised, notwithstanding that bird guide books would have them capitalised.

All other bird names are capitalised e.g. Northern Bald Ibis, Song Thrush, Herring Gull, including, for the sake of consistency to the reader, full capitalisation of hyphenated names e.g. Red-Legged Partridge, Edible-Nest Swiftlet, Rough-Faced Shag, notwithstanding that most bird guide books would have the latter displayed as e.g. Red-legged Partridge.

FLIGHTLESS BIRDS

EMU

Fatherhood – but not as we know it

Spare a sympathetic shudder for the hard-pressed male emu, the ultimate henpecked, cuckolded, devoted dad.

After the emu chicks are born, the father spends eight weeks incubating and protecting the eggs, with a fidelity that passes beyond the bounds of even his most attentive human counterpart. He hardly ever eats, drinks or even defecates, but survives by descending into a kind of stupor, letting his body temperature drop by 4°C to avoid losing liquid. Phew. Not even David Beckham, the unsurpassable New Man, can compete with that. Where is the mother, you are probably asking by this stage? The terrible truth is that she is roaming around another part of Australia looking for a new partner. Emus, like a small number of other species, are polyandrous, a scientific way of saying that it is the women who have all the fun – just like the female Red-Necked Phalarope, which enjoys Highland flings in the north of Scotland.

The modern male emu, staggering around after this two-month stint, can at least be thankful that life is not as bad as it was for his ancestors in the 1930s. Unlike them, he does not have to dodge bullets in addition to desisting from defecation. The emu is one of the few birds in history on which a country's army has ever declared war: annoyed by trampled crops, farmers persuaded the government to send an artillery squad to eradicate the emus of Western Australia.

This military escapade did not work out entirely according to plan, however, and the emus emerged victorious. These giant fowl may look bulky and unwieldy, but they proved surprisingly fleet-footed when necessity demanded. Emus can run at up to 30 mph.

They also proved unlikely masters of military tactics – adept at making strategic withdrawals in the face of overwhelming odds, only to regroup later. In World War Two, the Germans lost on the Russian front precisely because they proved unable to perfect this technique – but the emus managed it. After killing only twelve birds, the Royal Australian Artillery accepted defeat and called it a day. Nature cannot always be conquered by humanity.

But the male emu does not make life easy for himself, it must be said. While looking after the chicks he becomes extremely irritable and downright aggressive towards anyone regarded as an intruder – much like Rod Hull's Emu, the puppet famed for regular savage attacks on celebrities on television sets in the 1970s.

The emu also suffers from the same problem as other tasty birds in human society: it has attracted the attention of some characters with less than honourable intentions. While many other birds have been ignored by humans because they are bony or foul-tasting, emu meat is, by all accounts, similar to the best succulent beef (read more on the advantages of tasting disgusting in the Fulmar essay). This explains the rapid extinction of the Kangaroo Island Emu – scientifically considered a separate species from the sole surviving species, whose full name is the Spotted Emu. The eradication of the Kangaroo Island variety is blamed on Matthew Flinders, the explorer who landed his men on the island in 1802, at a time when the word 'sustainability' had not been coined, to find large hunks of fresh, flightless meat running around.

The emu's notorious inquisitiveness about humans would have done little to aid its survival on Kangaroo Island. Some bird species, like the scarce quail in Britain, are rarely seen because they are so shy, but others, like the emu, or the nightjar of England's heathland, will boldly approach humans to see what they are doing. We are not the only creatures to have an interest in the world about us that is unrelated to the simple daily struggle of finding enough food, and that is a heartening reminder that we have more in common with other animals than we sometimes think. However, just as curiosity killed the cat, it also put paid to the Kangaroo Island Emu – though not, thank goodness, to the Spotted Emu, which later learned that discretion was the better part of valour in its skirmishes with the Royal Australian Artillery.

So, all you new fathers, next time you're down the pub enjoying a guilty glass while mum is at home with baby, pray raise your pints to that paternalistic *primus inter pares*, the male emu.

KIWI

It was the worst of times, it was the worst of times

Kiwis seem the least bird-like of all birds. Imagine a kiwi without its bill, and concentrate instead on its fleecy feathers that resemble a mammal's hair and its lack of a tail. It looks more like a mole than a bird.

But the resemblance to mammals is not merely superficial – there are many similarities of physiology and lifestyle.

Most famously, kiwis can't fly. They have a great sense of smell and hearing, but a very poor sense of sight – like many mammals and most unlike birds, which can generally see better than humans. Kiwis can only spot things 2 feet in front of their rather long bills in broad daylight, though they can monitor developments 4 feet further in front of them at night, which sounds odd but makes sense when you consider they're nocturnal – like many land-based mammals. Their body temperature is 38°C, which is more like a mammal's than the average bird's 40°C or so. Kiwis also smell by probing with nostrils at the far end of their beaks, which sounds even odder but makes sense when you consider that, like many mammals, they use smell to find much of their food – their nostrils are nearer the food which they have to find.

The Brown Kiwi, the commonest member of the family, also sometimes looks after its young for long periods, like the more sophisticated mammals. In the Brown Kiwi's case, parents accompany their offspring for up to three years. This length of time is rare for birds, but makes sense for the Brown Kiwi, which produces a small number of young every year (usually just one or two). This creates both the time and the evolutionary necessity for it to help its offspring until they have become adept at the important skills of finding food and avoiding being eaten – particularly since Brown Kiwis take eighteen months to grow to full size, so are physically more vulnerable to predators for a long time. In this respect, their habits are much nearer to our own than those of most birds, which are quite ruthless about abandoning their young within a matter of months at most (read about the Golden Eagle's callous treatment of its fledglings in the Golden Eagle essay).

It is no mere chance that the kiwi's lifestyle has come to resemble that of mammals so much. After all, New Zealand, the birds' home, has no native

species of walking mammal – in common with many smallish islands that were cut off before mammals inhabiting the great land masses could colonise them on foot. It has a few bats, and whales and other marine mammals, but that's all.

But it was kiwis' adaptation to a New Zealand without mammals that very nearly proved their undoing. The trouble began when the Maoris arrived from Polynesia at some point before 1300. Maoris started proving a fact often glossed over by historians of imperialism – that westerners are not the only people capable of unsustainable levels of natural destruction, although their technology makes them the most adept at it.

Much of the Maori destruction of kiwis was an accidental by-product of another form of spoliation: kiwis' preferred habitat, tropical forest, was burned down to produce farmland. But Maoris also liked catching kiwis for food, and became ingenious at doing so – playing tricks on them like lighting branches so they looked like glow-worms, which kiwis like to eat, and hunting them with dogs. The Maoris were clever predators, but the kiwis were also easy prey, since they couldn't fly away, and had not adapted to run off very fast when danger approached, or to develop a healthy sense of weary cynicism when confronted with appetising-looking glow-worm-like objects. It is often pointed out that the Maoris regard kiwis as sacred, but in cultures across the world sacred status does not necessarily mean that birds are protected. Often it just makes the birds akin to chocolate – something you eat and enjoy, and then feel slightly guilty about afterwards. The Maori habit of roasting and offering the heart of the first kiwi to be killed to the forest god Tane to placate him for taking what they saw as his bird must have come as scant consolation to them.

Maori habits had wiped out one kiwi species, the Little Spotted Kiwi, from New Zealand's North Island before the westerners came in the nineteenth century. However, the arrival of European colonists sped up the process of destruction, since they hunted kiwis and laid waste to their habitat with industrial efficiency. The decision in the early twentieth century to start protecting kiwis saved the world's three or so species (scientists can never agree on the exact number), but the birds themselves deserve credit for their surprising adaptability in the face of adverse circumstances. Forced to find new habitat because tropical forest was disappearing, they responded by moving to temperate forest, scrubland, and even commercial pine plantations. They wouldn't have been seen dead in these places a few hundred years before because they don't provide the sort of dense cover that kiwis prefer, but given that tropical forests were being burnt down, they would only have been seen dead in them if they'd remained in place.

Kiwis are also admirably persistent about breeding, like many birds. If their eggs are taken, they will lay up to four replacement clutches before giving up for the year. This has proved handy in a country where their eggs are often taken by stoats. Stoats? What on earth are they doing, fraying kiwi nerves in a country that has no native land mammals? They aren't supposed to be there but were introduced by humans to keep down the population of rabbits, who aren't supposed to be there either, but were introduced by humans too. One imagines the kiwi's rage slowly building as its eggs are nicked. But it suppresses its anger and just gets on with the job of laying more eggs. To borrow an adjective inspired by one of the kiwi's bitterest enemies, it is an admirably dogged bird.

INACCESSIBLE ISLAND RAIL

When it pays to be hard to get

It may not surprise you to learn that few people have seen the Inaccessible Island Rail.

There are two main reasons for this, and both are given away in the bird's name. Most rails are small shy birds that skulk in the undergrowth, much like Britain's Water Rail. Moreover, this particular one can only be found on Inaccessible Island. Some place names are famously inapt – like Greenland, or the storm-ridden Cape of Good Hope. But the sailor, now forgotten by history, who named the island was a good geographer. It is miles away from Tristan da Cunha, the nearest island that anyone has heard of, and that is miles away from anything else. Should you manage to reach Inaccessible Island, to make it past the beach you have to scale high cliffs in your attempt to find rather a drab-looking brown bird that does its best not to be seen by people who generally are not there anyway – not, at least, since the Stoltenhoff brothers arrived from Germany in the 1870s to attempt to make a living by catching seals and trading their meat and fur, without stopping to think about the crucial flaw in their plan that there was no one to trade *with*. This was a rare but instructive case where a bird's name hinted heavily at the project's commercial viability. For more than a hundred years since then, no one has tried to live there permanently.

It is a pity, though, that the rail lives in a place that is so, well, inaccessible, since it is a fascinating creature scientifically. As the smallest surviving flightless bird in the world, it is only 17 centimetres from head to rather stumpy tail, making it even more diminutive than a skylark. It is a fascinating example of island dwarfism, though the even smaller Laysan Rail of the eponymous Hawaiian island, which only became extinct within photographic memory, was little bigger than a sparrow.

Island dwarfism happens when birds that live on small islands become smaller than other birds of the same species that live on large land masses. One possibility is that resources are limited, so only the smaller individuals tend to survive. Through a process of natural selection eventually all the birds of that species on that island are small. Eventually those birds on the island evolve into a separate species that can't mate (or, more strictly speaking, can't consistently produce fertile offspring by mating) with the related birds elsewhere, even if they were let loose on the mainland. Island dwarfism has a direct opposite, island gigantism, also caused by isolation – only in this case the birds get bigger, usually because they are filling a niche left vacant by the absence of large mammals.

The topic of island dwarfism had previously been a mere scientific curiosity that we had noticed in other creatures. But it suddenly became hot news with a direct bearing on us humans in 2003 when scientists declared they had found skeletons of a species of dwarf – related to *Homo sapiens* but distinctly different – on the Indonesian island of Flores. The newspapers imaginatively referred to these newly found creatures as 'hobbits', after the small human-like creatures in *The Lord of the Rings*.

Scientists still can't concur on whether Flores Man is actually a separate species, a mere sub-species (different from *Homo sapiens* but able still to mate with us), or simply a bunch of exceptionally short people who were fully signed-up members of *Homo sapiens*. Confusion deepened when it was pointed out that a village of exceptionally short people existed within walking distance of where the skeletons were found. Whatever the outcome of the debate over the Flores hobbit, the case of the Inaccessible Island Rail underlines the fact that a group of creatures can, if cut off from the rest of their species, grow progressively smaller and eventually evolve into something else entirely.

The irony is that if we could easily have seen the Inaccessible Island Rail, we might not be able to see it anymore. Many flightless rails in less out-of-reach places have become extinct because of the introduction of dogs, cats and rats that can readily eat them or the eggs they leave in nests on the ground. If there had been such a species as an Accessible Island Rail (which sounds more like a public transport service for senior citizens than a bird name), it might well have become as dead as the dodo – a flightless example, incidentally, of island gigantism.

DODO

Only the public can make a star

In the pantheon of defunct birds, the dodo holds the role of the *Big Brother* housemate that defied all the rules of celebrity to become a star. It was physically unattractive, graceless and looked so peculiar that it resembled a bird dreamt up for a cartoon rather than a real species. But despite dying out in about 1662, less than a hundred years after Europeans had the good fortune to discover it, and it had the ill fortune to discover us, on its island home of Mauritius, the dodo has lived on in our culture – even though far more beautiful birds have been forgotten by history after their disappearance. Who now remembers the stately Choiseul Crested Pigeon of the Solomon Islands, its gorgeous head bedecked with a blue fan that would have looked good on a ladies' hat – a cosmetic touch that contributed to its extinction some time in the twentieth century? The Choiseul Crested Pigeon is dead, but the dodo lives on.

This is all the more amazing since the dodo had the worst possible beginning on its road to becoming a posthumous star. A century after its extermination by the exotic combination of introduced rats and macaque monkeys, Linnaeus, the Swedish naturalist, added insult to injury by assigning it the Latin name

Didus ineptus – stupid dodo. He may only have been following precedent, since the word 'dodo' itself very possibly comes from a Portuguese word for an idiot – although there's also an intriguing suggestion that it may have derived from the dodo's call, given that taxonomists assume it was a kind of dove, and 'dodo' is not a bad rendition of the two-note cooing of many birds in this family. Think, for example, of the call of the Collared Dove, commonly heard in British gardens: 'Do-Doh-Do'.

Things went from bad to worse for the extirpated dodo. It couldn't even survive as an ex-dodo – the last known stuffed specimen, originally in Oxford's Ashmolean Museum, suffered severe decay, and little of it is left. By the nineteenth century, many writers thought it had never existed at all, but was a made-up bird – an amusing figment of the imagination.

But slowly the reputation of this flightless bird began to ascend in a lark-like manner. In his classic 1865 children's book, *Alice in Wonderland*, Lewis Carroll made it one of the more sympathetic characters – doubtless because

the dodo was a caricature of himself ('Do' for 'Dodgson', his real surname). Naturalists pointed out that it cannot have been a stupid bird if it had managed to survive millions of years on Mauritius while many other species became extinct around the world because they could not adapt to their habitats. It was instead, they correctly argued, a victim of humanity's fecklessness in letting predatory creatures loose on the island.

To boot, scientists argued that Linnaeus' low opinion of the dodo was probably based on drawings of it that showed an unusually corpulent bird, which appeared incapable of running anywhere and fleeing anything. Historians have long suspected that this is an inaccurate representation of the dodo in its natural state, arguing that the drawings were of the bird as a captive that had been fattened up for the pot – though since some accounts describe it as unusually disgusting to eat, dodo-chomping sailors must have been pretty desperate.

Now the dodo is held in curious affection, in a way that other deceased birds are not. The related dove that circled somewhere near the top of so many religious paintings of centuries past, symbolising the Holy Spirit, has lost much of its symbolic power, drained of potency by the decline of Christianity in the West. But its cousin the dodo goes from strength to strength.

Nevertheless, I have to confess that I still find the dodo a comical-looking creature, despite seeing recent reconstructions that show a leaner, fitter, altogether better bird – based on research unearthed in the past few decades, including newly found skeletons and rediscovered drawings made before the dodo's death. It is the bony bill that sets me chuckling, tapering from a bony head, only to end in a bulbous hook. It is a resoundingly ugly bird, though a fascinating one to look at precisely because it conforms so little with our image of what a bird should look like.

Did the dodo die in vain? Not if we can work out the magic ingredient that makes people care about its extinction. For the dodo, the answer clearly lies not in any sense of beauty, but rather in the idiosyncrasies of the creature. It would have been a fun bird to go and see, or at least to watch on wildlife programmes. One can imagine it being the star of an International Springwatch on the BBC.

The dodo would, were it alive today, be amazed by the change in humanity's behaviour towards rare new birds discovered on out-of-the-way islands. Back in the sixteenth century they were thought of as, quite literally, fair game for sailors. But nowadays scientists sometimes even baulk at the time-honoured ritual of hunting the holotype – killing a single specimen of a bird new to science to get the perfect description of every possible plumage detail and anatomical point of interest. Instead ornithologists often instead choose merely to observe the bird's habits closely, before briefly catching it to take a DNA sample and several photos. Killing rare birds without any thought before stuffing them for the pot or museum collection is now an almost extinct practice – but not quite as dead as the dodo.

SEABIRDS

ARCTIC TERN

Fly me to the moon

The thorny question of whether birds migrate bothered learned people for centuries until they finally reached agreement, in the early nineteenth century, that even small birds do after all engage in the amazing feat of flying hundreds or thousands of miles away to warmer climes where the food is more plentiful.

Until that time, the topic of migration proved to be one of the great banana skins among the world of scholars, constantly pushing great thinkers into intellectual pratfalls. The subject fooled men as distinguished as Aristotle, who concluded that, while some birds migrated, others such as the swallow went into a kind of hibernation. Gilbert White, the eighteenth-century English parson remembered by history as the world's first birdwatcher (for more on White's adventures, see the Chiffchaff essay), held the same view – that while some species flew south for the winter, many, though perhaps not all, swallows chose to 'lay themselves up like insects or bats, in a torpid state'. Other theories about the swallow's winter whereabouts suggested that they spent the season curled up in the bottom of ponds, or even flew to the moon. Some leading eighteenth-century scientists, such as the English polymath Daines Barrington, denied that any species of bird migrated.

It is easy, in retrospect, to ask how such distinguished men, who were right about so many other phenomena in the natural world, could have been so astoundingly stupid. Aristotle correctly suggested, for example, that larger birds tended to live longer than smaller birds – a fact not proved beyond doubt until ringed birds were studied in the twentieth century. If he guessed this, how could he have been so wrong about migration?

But in history, hindsight is both a wonderful thing and an unfair judge – and the history of birdwatching is no exception. Many of the things done by birds seem so outlandish that we would be stupid to believe them unless we had clear proof. Whilst people could see with their own eyes, without modern optical equipment, that storks migrated, firm evidence that smaller birds did so took much longer to appear.

A perfect example is the Arctic Tern, a white seabird with a forked tail, thin wings and a sharp beak for stabbing fish, which flies more than any other migrant. Some individuals breed in Greenland during our summer, and then spend our winter (the southern hemisphere's summer) on the edge of the Antarctic ice. Given that an Arctic Tern can live for twenty years, it can put in more than 500,000 miles of flying. This is easily enough to take it to the moon and back, so was it really so preposterous to have believed once that swallows did actually migrate there?

It is simple, armed with the knowledge of Arctic Terns' travels, to see that it all makes sense because they gain the advantage of two summers in one year – and who among us would turn that down if they had the opportunity? Because of their wanderings, Arctic Terns spend more of their lives in daylight than any other creature on earth. This doesn't just sound nice, it's useful too, since it maximises the daytime hours during which they can fish.

Another example of a bird with a migratory pattern that beggars belief is the Blackpoll Warbler, a pretty streaked creature only 13 centimetres long whose male has a clear white face with a black crown in its breeding plumage. The bird launches itself off the Massachusetts coast and into the Atlantic at the end of summer in an effort to get to South America, rather than taking the ostensibly easier way of sticking to land by flying through Central America.

The Blackpoll Warbler appears to be taking a suicidal step – how could such a small bird manage to fly without food or rest over such a large expanse of sea? But it makes perfect sense. By leaving the coast where they do, Blackpoll Warblers can take advantage of the powerful trade winds that more or less blow them to South America in only four days. So it's like being on one of those conveyor belts at an airport, or indeed on the plane itself. The strange thing is that natural selection has, over millions of years, rewarded those Blackpoll Warblers that seemed to their colleagues to be doing the dumbest thing of all, by enabling them to survive their venture over the ocean.

This leads to a knotty conundrum – why isn't the Arctic Tern called the Antarctic Tern, since birds that breed near the North Pole spend as much time near the South Pole? The most obvious answer is that there's a close relative called the Antarctic Tern already, which raises its young down south. If you delve into bird names you open a can of worms – which is of course good news if you're a robin. You would never find a Dartford Warbler in Dartford for example. Even more confusingly, a Red Phalarope in the US turns into a Grey Phalarope in the UK – because it's rarely seen in its bright summer plumage in Britain – and the Invisible Rail of Indonesia is, surprise surprise, not really invisible (it just seems so because it spends all its time sneaking around in dense vegetation).

The true travels of the Circumpolar Tern – to give it a new, more appropriate, and impressively august name – would stretch the imagination of the reader of a picaresque novel. When truth is stranger than fiction, one cannot blame scholars for believing in fiction. A general warning is in order: historical figures in the rear-view mirror are not as barmy as they first appear.

GUANAY CORMORANT

Where there's muck, there's brass

The Guanay Cormorant – named after 'guano', the Spanish for bird dirt – owed its more glamorous name as 'the billion-dollar bird' to the immense value of its excrement.

Where there is enough bird excrement, there is money, and where there is money there is always conflict. The Guanay Cormorant and its fellow seabirds that lived on rocky islets a little way off the western coast of South America were implicated in the outbreak of not one but two wars in the nineteenth century. It wasn't, of course, the Guanay Cormorant's fault. It was, as usual, a case of every prospect pleasing – particularly the prospect of huge piles of bird dirt – while only man was vile.

Bird dirt is an effective fertiliser, but the Guanay Cormorant does not produce particularly large amounts of excrement for its size. Its unusual value, for a brief period while stocks lasted, was based instead on two happy accidents. One was the fact that it lived cheek by jowl with other Guanay Cormorants in extremely tightly packed colonies that harboured hundreds of thousands of birds. The second was the dry climate in which it dwelt, which preserved the guano. There is also the advantage that most bird faeces are very thick and solid, because birds do not like losing water from their bodies if they can help it, since it must then be replaced by drinking. Because of all these factors, guano, enriched by other detritus such as bird corpses, had built up over thousands of years to layers up to 90 metres thick. Ninety metres definitely qualifies as a big pile of, umm, shimmering potential wealth.

Guano transformed the economies and societies of Peru and Bolivia, and later of Chile after it acquired Bolivia's guano riches. This natural resource was so important that the period from 1845 to 1866, when Peru saw strong economic growth thanks to guano under an unusually stable period of government, is still known with affection as the Guano Era – much as many (though by no means all) Britons look back fondly to the Victorian era that marked the acme of British power. Huge personal fortunes were made – though, as is usually the way, the actual people standing on cliffs and collecting the guano did not amass them. On Peru's Chincha Islands, one of the key locations for guano-collecting, initially convicts and slaves were used. By one of those strange quirks of human migration that has seen people from far away turn up in the unlikeliest places to do the unlikeliest things, the business of extracting the guano was soon taken up by Chinese immigrants, whose descendants remain in the country. They must, at some point, have scratched their heads and said to themselves, 'So by what accident of history did I end up doing this?' – though surely it's only a matter of time before fading stars on *I'm a Celebrity... Get Me Out of Here!* are given the same task.

It was guano that largely caused the Chincha Islands War of 1864–6. The conflict began when Spain tried to wrest control of the islands, which accounted for more than half the Peruvian government's revenues. It didn't, of course, tell the Peruvian government that it was declaring war over bird dirt. That would have been embarrassing. Instead it used the all-too-common pretext for a *casus belli*, alleged maltreatment of its citizens inside the defending country – that Peru had treated them like guano, in other words. Peru, supported by Chile, defeated the Spaniards, but not before Spain had considered consulting the British on exchanging the rich prize of the Chincha

Islands for Gibraltar. It would have been a lucrative but rather dirty jewel for the British Crown, and one wonders what the grey men of Whitehall would have made of this unusual offer, pondering the issue as they walked past the Westminster statues of imperial heroes covered, with a grim inevitability, in pigeon guano.

The second conflict – the 1879–84 War of the Pacific, aka The Guano War – was fought by Bolivia and Peru to defend their deposits against Chile's avaricious designs. A victorious Chile seized control of Bolivia's coast, and hence its guano – as well as the other benefits of having a coastline, such as ease of trade. Bolivia's lack of a coastline remains an emotive political issue to this day in the country, even though all the accumulated deposits have disappeared. The guano has gone, but the grievance remains.

Historians have argued that the sustainable collection of bird dirt would have been economically better for all the countries concerned, rather than the rapid and intensive plunder that happened. That would have meant practising what is done now – collecting guano in short bursts every one or two years outside the breeding season, so that the birds are not seriously disturbed and can produce new generations of guano-producing offspring. On the other hand, guano has become economically less valuable as new synthetic fertilisers have been created, so economists can argue that bird dirt has had its day and that rapid short-term pillage made sense, seen purely from a moneymaking point of view that doesn't take into account the welfare of the birds.

The Guanay Cormorant, which little knew what deep, er, trouble its excrement would land it in, saw a sharp fall in numbers because of the breeding disruption caused by year-round collection during the peak of guano mania – and who wouldn't be put off a little bit by a bunch of strangers crawling all over where you live, trying to dig out your excrement? Naturalists now think there are only about three million birds left, which is about a tenth of the number that prevailed in the nineteenth century.

The threat to the Guanay Cormorant now comes from the lack of available material at the other end of the digestive system. Scientists say it will be hard for numbers to recover because of the new threat of over-exploitation by trawler crews of the fish the cormorants eat. On the plus side, some experts have argued that current guano-collecting is good for the birds, since they are less likely to catch diseases and be plagued by parasites. What holds true for humans also holds true for birds.

ROYAL ALBATROSS

It don't mean a thing if it ain't got that ring

How long do birds live?

This is a question we could not possibly have hoped to answer before it was made possible by the modern practice of ringing birds, begun in 1899 by history's most famous Danish schoolteacher, Hans Christian Cornelius Mortensen. He began with European Starlings, but in the succeeding years thousands of species have been ringed to identify them as individuals whose life history can be checked. Larger birds are usually ringed with serial numbers, and smaller ones often with combinations of colours in different orders that make each one recognisable. Other more unusual methods have also been tried – one US ringer used biblical quotations. Who teacheth us more than the beasts of the earth and maketh us wiser than the fowls of heaven – particularly if we can get good information from ringing them?

You might ask how on earth a human can catch a bird in the first place, in order to ring it. The short answer is, with difficulty. The longer answer is, with an evermore ingenious selection of nets, including the Heligoland trap, a series of funnels made of netting which is easy to fly into but surprisingly hard to fly out of. It was first used on the eponymous German island by Hugo Weigold, the German zoologist. Weigold was also, incidentally, the first westerner to see a giant panda in the wild, less than a hundred years ago.

Ringing threw up some surprising findings. One of them was that many birds travel a lot more than we thought they did. We used to believe they moved great distances only for migration, but have since discovered, for example, that an albatross will casually travel a thousand miles in one direction as it follows the best feeding opportunities, and then a thousand miles the other way.

But the most surprising thing we discovered is how short a time most species live, and how long a time a small minority of other species do.

Ornithologist David Lack gave the British public a shock in 1943 with his revelation about the robin. It is one of Britain's best-loved birds, treasured most of all because it seems such a faithful friend, turning up in our gardens year after year, standing on the gardener's spade keeping him or her company, flitting around with an apparent sense of *joie de vivre* that cheers us up.

But in his tome *The Life of the Robin*, Lack pointed out that it was not the same bird turning up year after year, because last year's bird was most probably dead. Instead, what we were seeing was a succession of robins. The public was shocked and initially disbelieving. People would have been happier if their illusion of robins' longevity had been maintained, like the child whose canary seems to live forever because it is secretly replaced by the parents. But the truth was out: most songbirds live only for a year or two at most. A study of more than 10,000 ringed goldcrests – another small songbird – found that the longest-lived was only four years, ten months and nine days. Ringing and other studies also show that, even for those birds capable of living much longer, such as some seabirds, the bulk died in their first few years before becoming adults.

In evolutionary terms this makes good sense, though evolution is a cruel system. Only the strongest birds will survive for long enough to reproduce, and they will pass on the strongest genes. Small songbirds have the shortest lives but produce enormous numbers of young every year, so the mathematics makes this system tenable.

But what ringing also revealed was that, conversely, some ocean-going birds survived for much longer than we had guessed. Ignoring birds in captivity, which can live the longest of all because they lead riskless existences, one of the oldest wild birds ever recorded was a Royal Albatross, at fifty-eight. It was also, incidentally, still breeding at this age. In fact, looking at the numbers of albatross longevity, some scientists have speculated that the mortality rate (the proportion that die every year) among adults is so low that there should be a few birds that are over eighty.

Many seabirds are particularly capable of longevity. A Manx Shearwater – an ocean lover with long, straight wings that distinctively glides very close to the waves, with only the occasional burst of flapping – was identified and ringed as an adult of the species on Bardsey Island in Wales in 1957, and retrapped in 2003. This made it over fifty, since Manx Shearwaters become adults at the age of five or six. Ringing evidence also suggests that gannets – cream-and-white seabirds shaped like flying cigars – can live an extraordinarily long time.

But I can't help feeling a very unscientific soupçon of surprise that it's often seabirds that live so long. It sounds so dull, to fly the empty oceans for decades – a monotony interrupted only by bouts of breeding on some sparse rocky outcrop in the middle of nowhere. I am reminded of the legend of the Flying Dutchman, the ghost ship doomed to sail the high seas forever, and the Ancient Mariner in Samuel Taylor Coleridge's fantastical 'Rime' of the same name, forced to wander the world as penance for bringing doom to his ship by killing an albatross – used as a symbol for Christ in the poem because its long body and even longer, rather straight wings give it a cruciform appearance. What Coleridge could never have imagined at the time, in the era before ringing, was that the albatross might have been more ancient than the mariner.

FULMAR

A bird to die for

Can birdwatching actually be dangerous?

Birds can be extremely aggressive when defending their territory, as anyone who has encountered an irate fulmar can tell you. These grey-and-white seabirds, which everyone thinks are gulls but are actually petrels, look rather attractive from a distance. They can be identified by their long thin wings, held very straight, with a speck of white on them that looks like a splash of paint. But they are willing to attack an intruder with foul-smelling oil which, if spat onto your coat, cannot be removed even by the most innovative dry-cleaner. People who work with seabirds say that in these cases the jacket is always a write-off and best thrown away. This oil also has the protective characteristic of making fulmar meat disgusting to everyone apart from the erstwhile inhabitants of Scotland's St Kilda archipelago, now abandoned by all but the military and a few summer scientists. They thought it was delicious because they had grown up with it – making fulmar meat the Marmite of the bird world. History does not record whether the St Kildans smelt funny as a result of their diet, though as they lived on a remote island chain where they all ate the same smelly bird, this presumably would not have been a severely socially debilitating problem.

Fulmars are not the only birds that attack humans. Terns sometimes even team up with each other in colonies to mob a human intruder.

But does birdwatching ever cross the line from unpleasantness to danger? Eric Hosking, the pioneering English bird photographer who died in 1991, lost an eye and could have lost his life after he got too close to one of his subjects, an angry Tawny Owl looking after its young. He later called his autobiography *An Eye for a Bird* – the first recorded case in which this expression had a literal meaning.

Can it get even more dangerous outside Britain? It was discovered recently that the Hooded Pitohui, a black-and-orange songbird in New Guinea, has poisonous plumage, caused by eating certain kinds of beetle.

But the most dangerous bird in the world is the cassowary of Australia and New Guinea. It has three toes on each foot, with one toe on each bearing a nasty sharp claw that can cut people open. There is even a well-documented record of a cassowary killing a sixteen-year-old boy. To be fair to the cassowary, the boy and his brother were in fact trying to kill the bird – so it served the brat right.

However, the greatest danger to birdwatchers comes not from the birds but from themselves. When birdwatching becomes an obsession, people can take risks that don't pay off. A professional British birder called David Hunt was, like every naturalist who goes to India, determined to see a wild tiger. The ambition is rarely realised, but Hunt had the good, or bad, luck to see one. The strange thing is that when Hunt saw the tiger, he kept shooting photos of it as it came closer and closer, rather than running away. The last picture, developed after Hunt's death, is of the tiger's snarling face filling the whole shot, about to kill him.

It is often men who take the most extreme risks in birdwatching – though this is partly because a disproportionate amount of birdwatchers are male. Two young men were killed by Shining Path guerrillas in Peru despite being warned by villagers not to go into the rebels' territory for the sake of spotting birds. Another young birdwatcher died alone of dehydration in the

Australian bush. Some have lost their lives for a particular prized bird – like the ornithologist who headed out late one stormy afternoon in the Himalayas because he heard the call of that much celebrated species, the Satyr Tragopan – a gorgeous red pheasant with a blue head and clear white spots over its belly, as if it has just been crying and the tears have stained its chest. Perhaps the most beautiful birds are ultimately the most dangerous ones, because people take such risks to see them.

But the biggest risks of all are taken by people who want to see all the birds on this earth, however drab or beautiful. These are the world listers, who devote their lives to seeing as many species as they can. Several have died while on the job. They rarely do anything foolhardy, but many years travelling in small planes to land on small airstrips, or on bad roads in pursuit of an elusive endemic – a bird seen in one area and nowhere else in the world, such as the Scottish Crossbill – add up to a statistically significant risk of coming to an untimely end. Some of these people eventually pay the price for their habit of going to dangerous places in dangerous ways year after year. The most famous of all, Phoebe Snetsinger, had used the fortune inherited from her father, the advertising magnate Leo Burnett, to become a world lister. She died in a traffic accident in Madagascar in 1999 while pursuing her dream. Ironically, her world listing had been a response to what she thought was her impending death, as she took it up in 1981 after being told she had terminal cancer. Snetsinger decided it was better to die while doing something she loved – and so she did, though not in the way she had expected (for more on the strange phenomenon of world listing, see the Red-Legged Partridge essay).

GREAT SKUA

Pirate of the high seas

The skua might sound the ideal house guest, for it will eat almost anything you put in front of it. Starting with a dainty compote of small berries, it will happily move on to a selection of small rodents and then a main course of fish. But unfortunately for any avian host, it does not stop there. It will happily eat your eggs, your children (even if you're a fellow skua), and ultimately yourself – often by seizing hold of your body and drowning you. Even if you manage to survive, it will probably nick your food.

Skuas – rather thuggish-looking, gull-shaped birds with short, thick muscular bulldog necks and large hooked bills – are the paramount example of many birds' ability to adapt to whatever food happens to be there. It is grimly appropriate that in the 1930s the British Fleet Air Arm decided to name a deadly dive-bomber-cum-fighter-plane, the Blackburn Skua, after the bird. It was a Skua (rather than a skua) that downed the first Axis plane in World War Two.

One of the most common questions I'm asked, when someone sees a bird while out walking with me, is 'What does it eat?' Some small birds are picky consumers, whose bills have evolved to allow them to feed on a particular source of food. But many larger birds have the wherewithal to be more open-minded. Larger birds of prey have even been known to supplement their usual diet of songbirds and small mammals with the odd smaller bird of prey, if they get into a brawl with one over territory and happen to kill it.

Skuas are among the least fussy eaters of all. Like the perfect tourist eager to sample the culinary fare of wherever he or she is staying, they will take to the local food, whether lemmings, fish or fellow seabirds. But food-stealing is the habit for which skuas have become best known. The posh scientific name for this is kleptoparasitism, which, roughly speaking, is Ancient Greek for stealing by eating from somebody else's table. Many scientists say, more romantically, that skuas are 'piratical', and this neatly sums up their chosen way of life. Like human pirates, they will seize other birds' hard-won treasure – forcing seabirds that have just caught fish to disgorge it from their throat or even from their stomach. In bird as in human life, threatening behaviour is usually enough to get what you want without having to resort to physical violence, but skuas are willing to lock their beaks onto birds' wings and drag them down into the sea until they surrender their food. But although skuas are kleptoparasites, they are not kleptoparasitical coprophagiacs, as people used to think they were in previous centuries. This sounds impossibly grand, but it means chasing after other creatures to eat their poo.

To be fair to skuas, other birds demand food with menace too. Such thuggish behaviour – rather than outright hunting and killing – is quite common among birds in general. Blackbirds, in particular, like stealing snails from other birds with the threat of a little GBH (after the shells have been conveniently smashed of course), so there are probably pirates at the bottom of your garden. Shiver me timbers. Writers have been prone to hold up birds as an example of ordered existence that we badly behaved humans can learn from – and thinkers have even made a case for saying that human violence must arise from the corruptions wrought by society, rather than from human nature itself, since birds and other species behave so well to creatures that are not their direct prey. But in truth birds are not so very different from humans living in a rather lawless country, where many individuals are prone to steal from each other, and not afraid to use violence to do so. Birds can also be amazingly opportunistic in what they eat – many years ago there was a long correspondence in the letters of *British Birds*, the scientific magazine, on the diet of the turnstone (a wading bird about the size of a Great Spotted Woodpecker, found on the coastline throughout Britain in the winter), which culminated in a reader's observation that he had seen a turnstone eating a human corpse washed up on the beach.

The varied diet of skuas perhaps explains another peculiarity shared with only a small number of other birds. Some skuas of the same species have very dark plumage, whereas some are largely yellow-and-white. The 'dark morph' birds make for better pirates since it is harder to see them coming than it is to see a 'light morph' bird – and skua populations that are more inclined to banditry during the crucial breeding season, when they need as much food as they can get whether by fair means or foul, have a higher proportion of dark morph birds.

Scientists have, in the past, found these different morphs terribly confusing – and none more so than those of the Arctic Skua, which even has the bad manners to have an 'intermediate morph' not shared by other skuas. When you add to this the fact that the juveniles look very different from the adults, you have the perfect conditions for confusion. Early naturalists, who naturally liked finding new species even when there were none, took full advantage of this, to the point where sussing out new skuas seems to have become a bit of a cottage industry. A nineteenth-century expert on them looked at no fewer than twenty-three different skua species named by naturalists over the years and decided they were all just one – the Arctic Skua, which like the Great Skua breeds in Scotland. Spoilsport.

Great Skuas didn't arrive in Britain till they drifted down from the far north to reach Scotland in about 1750, but they have gradually prospered precisely because they are so good at taking food from us – although they are not really stealing from humans, because they tend to feed on fish discarded as unsuitable by our fishing boats. Conservationists' principal worry is not that there are too few of them in Britain but too many. If fishing policy becomes stricter and there are fewer fishing boats to steal from, the Great Skuas could rely more on their old habits of purloining food from other seabirds, which could in turn starve as a result. Here is perfect proof that changes in human behaviour tend to set off a chain reaction in the world of birds, the eventual consequences of which are often wide-ranging and highly unpredictable.

EMPEROR PENGUIN

A history of hot water

For a bird that, for the most part, prefers the cold nethermost regions of the world, the penguin has found itself in an enormous amount of hot water at various points in history.

Until the birds were discovered by Western explorers about half a millennium ago, they had led a harsh though uncomplicated existence.

The Emperor Penguin had the harshest time of them all. A huge, waddling creature with black wings, a white belly and a neck and throat tinged yellow, its image on the Penguin chocolate bars inspired the fondly remembered Pick-up-a-Penguin advertising catchphrase. You wouldn't actually want to make an attempt to shift an Emperor Penguin very far towards the skies though – adults are up to 4 feet tall and weigh in at up to 45 kilogrammes.

The creature on the chocolate wrapper is a tough cookie. Emperor Penguins are the only birds that take the masochistic step of breeding in the Antarctic winter, when temperatures can fall below –60°C and the wind reaches speeds of over 100 mph. Unsurprisingly, their shape and even their social life have largely evolved in reaction to the freezing conditions. Their short wings prevent the loss of body heat (think of how you huddle your arms close to your chest when you're cold, and you'll sense how much heat can be lost through the limbs). Emperors and other penguins also preserve heat by huddling together, providing a perfect illustration of the rule that sometimes in nature co-operation is better than competition. Many penguins nest in large colonies, constantly shuffling around so they can take turns at the edge of the huddle, where the wind is coldest and the greatest energy is lost. It is their huddling instinct, the Antarctic equivalent of the herding instinct, which led to perhaps the strangest ecological disaster in history – the trampling to death of 7,000 King Penguins on Macquarie Island, halfway between Australia and Antarctica, after an Australian Air Force plane came too close and set off a stampede by the densely packed birds.

Tough cookies attract each other: after Robert Falcon Scott set off in 1910 to reach the South Pole, three of his men went on an introductory jaunt to procure an Emperor Penguin's egg. The expedition, carried out in the middle of the Antarctic winter in the most appalling conditions, testified to the masochistic tendency among Britain's ruling classes to pick the hardest task imaginable and set about it in the hardest possible way, with the least possible fuss. When they returned with the egg, their upper lips well and truly frozen stiff, Scott told them that in his opinion it was the hardest journey ever undertaken by humanity.

The filching of an egg, however, was a small misdemeanour compared to what humans had got up to in previous centuries. Relations between the two sides had started badly hundreds of years before when we accidentally gave them the wrong name, calling them penguins (from the Latin *pinguis* for fat) because we thought they were Great Auks – fat, black-and-white seabirds from the auk family that became extinct in the nineteenth century.

Centuries of persecution ensued. Corpulent penguins, and their large eggs, made good meals for sailors, and the generous deposits of oil in their bodies could be used by sailors and locals alike for fuel, heating, leather-tanning and a host of other ingenious applications (although, of course, the generosity was accidental on the penguins' part).

By the early twentieth century the persecution had largely stopped, after some hard-fought battles by conservationists. But it was not long before penguins became embroiled in a new series of debates – ranging from the ethics of putting animals in zoos, to the controversial concept of the 'just war'. In the twenty-first century the Chinstrap Penguin, another species related to the Emperor, has even become an unlikely icon for gay rights.

Why have penguins, in particular, been mixed up in such things? The simple answer is probably because they look so much like people, as these unusually upright creatures walk around keeping up a constant chatter, like a man in a dinner jacket whose innate loquacity has been uncorked by the consumption of a few too many bottles of wine at a posh banquet.

This rather human appearance explains why penguins have become such popular exhibits for zoos – a fact that has made them a focal point for arguments about whether wild birds should be confined in them at all. Their anthropomorphic manner also explains why they turned into a regular fixture of *The Guardian* cartoonist Steve Bell's fulminations against Britain's re-invasion of the Falklands in 1982. The use of penguins, which live on the islands, initially emphasised how remote the Falklands were from their British overlords – and hence how absurd and unjustified the war was in the eyes of the campaign's opponents. Having discovered the human-like qualities of the bird, Bell later borrowed the penguin again, this time portraying a tubby penguin character as an agent of capitalist greed, who ran privatised prisons and a tabloid newspaper.

The humanoid penguin's latest wade into the deep waters of controversy came in 2005 with the publication of *And Tango Makes Three*, a children's book based on the real-life case of two male Chinstrap Penguins who formed a bond and raised a chick in New York's Central Park zoo. The book was popular with teachers trying to spread the message of gay rights, and thus equally unpopular with many American Christian parents. It became the most-banned book in US libraries, and the subject of a series of court cases debating the finer points of the Constitution's First Amendment, which guarantees free speech.

As always, truth was more complicated than fiction. One of the penguins turned out to be bisexual, leaving his partner of six years to mate with a female. He p-p-picked up a new penguin.

HERRING GULL

Bothering the bouncer

The Tinbergens were two brilliant Dutch brothers who both won Nobel prizes. But while Jan earned his in economics, Niko acquired his laurels in 1973 for a lifetime spent teasing birds, specialising in the mental torment of the hapless, harassed Herring Gull.

Niko Tinbergen's special interest was in working out birds' instinctive behaviour, and he spent a happy summer in 1937 frustrating geese with his Austrian friend Konrad Lorenz at Lorenz's home. Starting with the observation that when a goose's egg falls out of its nest it will stretch out its neck, tuck the egg under its chin, and put it back in with the other eggs, they made a huge fake egg for the goose to see how it would react. The mischievous pair of friends saw that the goose still obeyed its instinct by trying to put it back in the nest even though the egg was clearly too big for the bird to succeed (this may make birds look dumb, but to see just how clever they can be, see the Carrion Crow essay).

Surrounded by a devoted group of students that included such future household names as the scientist Richard Dawkins, Tinbergen carried out similar experiments on gulls, going to the extreme of spending a year in

Greenland devoted to studying them. He had a particular fascination with the Herring Gull, a chunky grey-and-white bird whose thick yellow beak is tipped with a single large red spot that looks rather like a drop of blood. The Herring Gull is often found sitting menacingly on sea walls, resembling a bouncer at the nightclub of a decaying British coastal town, though it has a cosier alter ego as the gull making the mewing sounds at the opening of *Desert Island Discs* on BBC Radio 4. Tinbergen's interest culminated in *The Herring Gull's World*, penned in 1953.

Tinbergen used the distinctive red beak-spot to conduct another famous experiment. Armed with the observation that Herring Gull chicks begged for food after seeing the red spot, he put in front of them a red knitting needle with white bands painted on it, to give it the appearance of an acne-ridden Herring Gull with a profusion of red spots, rather than the mere single spot that Herring Gulls have in reality. The chicks responded to this stimulus by

begging more eagerly than they did when confronted with an accurate model of the single-spotted parent gull.

The pattern repeated itself for other experiments. Tinbergen found that small songbirds preferred to sit on the largest artificial eggs possible, even if the eggs were so big that they slid off them. They also favoured bright blue eggs to their own plainer ones. Turning his attention to butterflies, he found that males preferred to mate with cardboard copies than with real females, if the cardboard copies mimicked the females' features in an exaggerated form.

Tinbergen used these experiments to coin the concept of 'supernormal stimuli': creatures react more to exaggerated versions of the real thing than to the real thing itself. The concept was taken up by psychologists and applied to humans – recently and notably by the Harvard psychologist Deirdre Barrett, who argued in her 2010 book *Supernormal Stimuli: How Primal Urges Overran their Evolutionary Purpose* that many of the problems of modern affluent societies are triggered by this principle. So, for example, we have a natural craving for fats because in primitive societies they prevented starvation, but this craving has been exploited by junk food makers' creations of ever sweeter, artificially created junk foods that overindulge our desire. The end result is obesity. Barrett has also argued that television is so hypnotically popular because it is a supernormal stimulus. Our senses respond strongly to laughter, smiling and sudden action in our everyday lives, and because TV offers more of this than reality does, it has captured the brain. Tinbergen's findings also influenced his friend John Bowlby, the distinguished psychiatrist who argued that the evolution of instinct had been underplayed in the study of human behaviour.

We do not have to be Harvard psychologists to see the presence of 'supernormal stimuli' in our everyday lives, such as the Page Three glamour models with outsized silicone breasts in tabloid papers, but at least our understanding of the charms of Samantha Fox has been enhanced by Tinbergen's observation of the Herring Gull.

In the end Niko Tinbergen shared the Nobel Prize in Physiology or Medicine, for his groundbreaking research into instinct and behaviour in birds, with his friend Lorenz, as well as with Karl von Fritsch, an Austrian more interested in the bees than the birds. Tinbergen was always captivated more by the behaviour of birds than of humans, but we have learned an enormous amount about ourselves from him. Socrates said, 'Man, know thyself, and thou shalt know the universe and the gods.' Perhaps he should have said 'Man, know thy birds, and thou shalt know thyself.'

BERMUDA PETREL

The last word in Lazarus birds

The Bermuda Petrel, a strange-looking black-and-white seabird with a hollow tube on top of its beak that allowed it to pick up the scent of food in the wind, died out in the 1620s despite one of the world's earliest conservation attempts by the governor of the British-ruled island. It had fallen victim to the usual motley mixture of rats, cats and dogs introduced by settlers, as well as colonists who were after a decent meal.

The extinction was another blemish on the record of the West, which couldn't seem to occupy territory without exterminating the locals. But then, in 1951, a funny thing happened. Bermuda Petrels were rediscovered by ornithologists on rocks close to the main island – eighteen pairs of them, corroborating a single specimen collected earlier in the 1900s which had looked suspiciously like a Bermuda Petrel. In the following years, the Fiji Petrel was rediscovered. Then the New Zealand Petrel. And, most recently, the Beck's Petrel. In fact, the only extinct petrel that hasn't turned up again (yet) is the Jamaica Petrel, which hasn't been seen since the 1870s.

Petrels seem particularly prone to become 'Lazarus species', jargon for birds long thought extinct, which return from the dead like the Lazarus of biblical fame. But the same has happened to plenty of other birds too. Tom Gullick, an Englishman who has seen the most bird species in the world (for a little more on his adventures, see the Red-Legged Partridge essay), has been lucky enough to rediscover two: the São Tomé Grosbeak and the Yellow-Throated Serin in Ethiopia.

The Bermuda Petrel is simply the most extreme example of Lazarus species among birds, with the longest time span – three centuries – between extinction and resurrection. In the mammal world, the Caspian Pony, a breed of horse, was rediscovered in the 1960s about 1,300 years after it was thought to have died out, suggesting that the Lazarus possibilities in the bird kingdom are also (almost) limitless.

Does this rather comic and chaotic state of affairs make a mockery of science, by suggesting that ornithologists don't really know what's out there and what isn't?

To a certain extent, these pre-mortem pronouncements are inevitable. Taking petrels as an example, they spend most of their life at sea, and most look pretty similar to each other – making it easy to overlook birds that you are not looking for anyway because they are supposed to be extinct. They also tend to breed on isolated rocks that are hard to find. At the time of writing, no one actually knows where the New Zealand Petrel nests. So finding scarce seabirds is like looking for a needle in a haystack – or, if you prefer, looking for a needle in a haystack is like looking for a rare petrel in a sea.

There are plenty of valid excuses for other such Lazarus species. Take the Jerdon's Courser – a bird related to the lapwing, the black-plumed, broad-winged bird of Britain's fields that dives and soars abruptly like a butterfly. The Jerdon's Courser lives in forests in India, and was rediscovered only in 1986 – almost a century after the last sighting. Scientists had assumed that it went about whatever business a Jerdon's Courser goes about during the day, like related birds – but in actual fact it is only active at night, so they could easily have been looking in the right place but at the wrong time.

But the phenomenon of the Lazarus species also raises an important question about the philosophy of science. Many scientists adhere to the 'precautionary principle' – the imperative to assume the worst, on the grounds that this will spur the greatest effort to improve the situation. They say this works for birds, and they can point out, for example, that many ornithological expeditions have been launched to find a bird precisely because they are thought extinct – think of the glory of rediscovering it. When I was at university a group of undergraduate birdwatchers went off to Madagascar for their summer holiday and rediscovered the Madagascar Serpent-Eagle. How's that for a student lark?

Most controversially, the precautionary principle has been cited to defend radical action against man-made climate change. Scientists have argued that by the time we have established with absolute certainty that the earth is perpetually warming up and that humanity is causing this, it will be too late to do anything about it, so we have to act now – on worst-case assumptions.

Why, then, when there are so many 'extinct' birds left to look for, do so many teams sail off to look for bizarre and spectacular creatures that have caught their imagination, but have clearly never existed, or have long since become extinct? Recent years have seen renewed searches for the yeti, the Loch Ness monster and even the Sabre-Toothed Tiger. One answer is the growth of the Internet, the world's most effective spreader of tall tales, which has given new life to the search for creatures whose existence is mocked by academic scientists – a fad known as cryptozoology. This is not so much the precautionary principle as the preposterous principle.

I have to finish this essay now though. I've just seen a funny-looking bird flying over the garden pond, and I think it might be a Jamaica Petrel.

ROUGH-FACED SHAG

A splitting headache for conservationists

Life is tough for the Rough-Faced Shag.

Until recent times New Zealand's two or so species of shag – black-and-white birds with big feet – were distributed across a wide range of islands. Numbers were not enormous, but were in the many thousands. Life was dull, but safe.

And then, suddenly, mad scientists started madly dividing and sub-dividing the two or so species (nobody ever agreed on exactly how many there were even before the dividing began) into many different species, as if there were no tomorrow. Now there might be no tomorrow for the Rough-Faced Shag – one of the rarer ones – because ornithologists have decided that after all the splitting, there are fewer than 300 bona fide specimens of the species left. To add insult to injury, the newly decreed species was given one of the most ridiculous names among birds – after the yellow-orange swellings found above the base of the bird's bill.

The Rough-Faced Shag illustrates a common new dilemma for naturalists – it is increasingly hard to preserve each species because the number of new species is increasing all the time. Only a minority of these birds was not previously known to science. The origin of the problem is that in the centuries-old battle between 'splitters' (who like dividing up a species into two or more) and 'lumpers' (who like doing the opposite), the splitters have won a spectacular victory. If the lumpers had their way there would be well under 9,000 species of birds, but now most scientists say there are about 10,000, and some put the figure at 12,000.

Why are there suddenly so many more species than previously thought? It is not because a lot more naturalists are going around measuring infertility – the key dividing line, since species that cross-breed with each other produce offspring that tend to be infertile. That's not a very feasible thing to do, so instead scientists have conventionally looked at differences in appearance between different birds. They argued that if birds looked pretty dissimilar, they were separate species, but if they only looked slightly different they were probably simply different races of the same species. However, the relatively recent science of DNA testing has allowed naturalists to discover new species by working out that the DNA of two very similar-looking birds is actually rather different. Hence the birth of the Rough-Faced Shag. And the Stewart Shag. And the Chatham Shag. And the Auckland Shag. And the Bounty Shag. And the Campbell Shag.

If there are a thousand more species because of DNA splitting, there are a thousand more species to conserve. Conservation is made yet harder by the fact that many of these 'new' species have a very restricted range. They are not rare because their numbers have fallen; they are rare because there were so few of them in the first place. The Rough-Faced Shag has only a few rocks in Marlborough Sounds, off New Zealand's South Island, to call its own.

Cynics might ask, do all these species need to be preserved? My answer would be that it is important to save all species, but that some species are more important than others. The most common argument used by scientists for conservation is biodiversity – that humanity benefits from the existence of as varied a set of creatures as possible. One such benefit of diversity is the pleasure gleaned from seeing such a wide range of species. This would place the importance of preserving a kagu, a beautiful grey-blue bird in the Pacific's New Caledonia island chain that looks very vaguely like a stork but is in fact not very much like any other bird in the world, above the priority of preserving the Rough-Faced Shag, which does have near-relatives.

Another benefit of diversity is that each different species may be useful to us – by giving us medical, industrial or lifestyle benefits that we can use, if we exploit them in a sustainable way. The mystery of the Lone Woman of San Nicolas illustrates both the virtues and the limits of this argument. The woman, who lived alone on the island off California in the nineteenth century after getting left behind during a slightly incompetent attempt to rescue her Native American tribe from aggressive Russian hunters, was eventually found after eighteen years, dressed in a skirt made of feathers from either the Brandt's or the Double-Crested Cormorant, which both nest on the island. The skirt was useful because cormorant feathers are semi-waterproof. But the feathers of any of the world's forty or so species of cormorant would have served – so the case of the mystery woman's clothing does not make a good argument for the preservation of individual cormorant species, such as the Rough-Faced Shag.

But we should return to the more important issue of what to do about the Rough-Faced Shag's silly name. On account of its large size, some nature lovers have started calling it the King Shag. That's much more dignified.

BIRDS OF PREY

PEREGRINE FALCON

Flash and fast

The fastest bird on the planet acted, half a century ago, as an early warning system to inform humanity that its mania for overusing harmful chemicals was likely in the end to hurt humanity itself.

The Peregrine Falcon is in fact not just the fastest bird but also the quickest creature in the world – more rapid than anything in the sea or on land, as well as in the sky. The highest recorded speed for the dive it uses to catch prey – 242 mph – makes the land-bound cheetah, which reaches 75 mph, seem a veritable sluggard.

The peregrine's speed made it the bird of choice for falconers. Generally speaking, common people had a kestrel (for more on this hovering wonder, see the Kestrel essay) and ladies had a merlin, a falcon considered suitable for dainty females since it was Britain's smallest bird of prey. But lords and other grand men had a peregrine, often kept in a special cage called a mew, since 'mew' is an old word for a falcon moulting, and since moulting falcons couldn't fly so well they were particularly prone to be kept in cages, in small buildings at the back of the main house. Just as the bird gave its name to the cage, the cage eventually gave its name to the building, putting the expression 'mews house' into the English language. After falconry all but died out they were often used for keeping horses, clouding the etymological origin of these buildings. Recent history has changed their role yet again: after the decline of the horse in towns, mews houses have become sought-after residences for just the sort of upper-crust people who, some centuries ago, would have kept peregrines.

The peregrine's athleticism has allowed it to perfect one of the flashiest of courtship rituals. The male drops food from the sky as an offering to the female, which flies underneath him to pick it up, briefly flipping upside down with perfect synchronisation to catch it in her talons.

It also exploits this nimbleness to catch mid-sized birds like pigeons, and the plethora of plump pigeons in British cities has allowed it to colonise

our most urban landscapes. I have seen this bird on an office block next to Regent's Park in London, with the adult keeping a proud and watchful eye on its successfully raised young as they stared curiously at the seething mass of humanity wending its way beneath them on the Marylebone Road. This love of cities might explain why the Air Ministry singled out the peregrine for persecution during World War Two – killing them before they polished off homing pigeons that were used to carry messages (in particular, a single pigeon was often put on a warplane, so that if the airplane was shot down, the pigeon could be sent out to alert the squadron to the airmen's plight).

The peregrine, superb aerial predator that it is, was the avian world's answer to the Spitfire at the time, so it was ironic that it had such a bad war in Britain, in contrast to other birds that positively benefited from the bombing and flooding that went on. However, all was not lost because *Falco peregrinus*, named after the Latin for 'wandering', was one of the most widespread birds in the world (the only non-icy land mass which it shuns is New Zealand, for reasons known only to itself). It therefore had an assured future around the globe, and even in Britain numbers had more or less recovered to pre-war levels by 1955.

But then scientists started noting something strange, in Britain and elsewhere. Peregrine numbers were declining, and for the oddest reason: adults were not producing young, despite forming pairs and finding territory in places rich with prey. This seemed to disobey the rules of nature.

Further research found a very unnatural explanation. A pesticide sprayed on crops, DDT, was building up in the peregrines' bodies, and reducing the amount of calcium in the eggshells laid by the females. Eggs were cracking before the young were ready to hatch. As a result, by 1963 only one in six pre-World War Two territories was producing young. Other birds of prey were being hit too.

Why was this happening to birds of prey, but not their prey itself? The answer was 'biomagnification': DDT was not accumulating to dangerous levels in smaller birds, but it was in raptors – birds of prey including falcons, eagles and owls – since they were eating hundreds of small birds and retaining the DDT in their bodies. The top of the food chain was suddenly the riskiest place to be.

Governments started forbidding the spraying of DDT on farmland, and this turned into a worldwide ban on agricultural use – though it is still employed to combat malaria. Peregrine numbers recovered across the world. By 1985, they were higher in Britain than pre-war levels.

The issue could have ended there, with the peregrine's future preserved and the case closed, but it did not. Humans started worrying whether we, at the top of our own food chain, might be affected too. *Silent Spring*, a groundbreaking book by US biologist Rachel Carson, attacked the overuse of DDT and other pesticides in 1962, and raised issues about their effect not just on birds but on *Homo sapiens*. The book became a key text of the emerging green movement. If birds have no future, do we? Perhaps not, but we do at least have an early warning system: the Peregrine Falcon.

RED KITE

Making the best of what you're given – or can take

To say the Red Kite prefers the remote Welsh woodlands is a bit like saying, a hundred years ago, that poor Italian immigrants to the United States preferred the overcrowded and insalubrious tenements of New York's Little Italy. In both cases, that was all that humanity allowed them.

In medieval times the Red Kite was possibly Britain's commonest bird of prey. It was familiar enough to give rise to the artificial 'kite' that children fly, whose technique for staying in the air – soaring, manoeuvred by its tail – is similar to that of the bird. Shakespeare constantly refers to it, and old descriptions of the bird mention its great numbers. However, these multitudes of kites were not in the isolated Welsh countryside where the bird had its last British toehold for most of the twentieth century, but in precisely the opposite environment: densely populated towns where, before binmen came and collected our rubbish every week, there were plenty of opportunities for kites to feed on the carrion. What more could they ask for? They didn't even have to overpower their prey, since it was dead already. Looking at it from the opposite point of view, what more could we ask for? In an age before organised rubbish collection by local government, Red Kites provided organised rubbish collection by birds. In reward for their free services, in Tudor London they were protected by law from being killed.

But as civilisation developed, kites lost their use, and before long fell into the very opposite category: that of a pest that might steal livestock, and was consequently worthy only of extermination. When modern firearms came along, Red Kites were quickly shot out of the sky. As a result, like other birds of prey such as the Golden Eagle, after a while they were restricted to areas where people and their animals didn't live. A crowded country did not leave many such places, so by the early part of the twentieth century they were down to about ten pairs in the woods of Wales.

The remaining Welsh kites soon became greatly loved by locals, and the birds were closely protected from egg collectors, who operated according to the grisly logic that the more endangered a bird, the rarer and more highly prized the egg. Welsh Nationalists saw them romantically, though with little sense of past history, as a symbol of Wales that required cherishing. Feasting on this goodwill, Red Kites were even forgiven their eccentric habit of pilfering unusual items such as underpants for their nests. However, the Welsh kites' success rate in bringing up their young to survive to adulthood was disappointingly low, so numbers recovered extremely gradually.

But in 1989, conservationists started introducing birds into the much more open countryside of the Chiltern Hills near London. What seemed foolhardy to some turned out to be a spectacular success. The kites have bred, proverbially speaking, like the rabbits that they often feed on. Scientists put two and two together and asked whether a Chiltern-style habitat was better for the kites than the wooded Welsh valleys.

The history of the kite in England shows that birds are what we allow them to be. If we decide that they are creatures of wild and lonely places, then they become so. That may be a sobering conclusion, but being reduced to living in a place which is not your ideal (what conservationists call a 'sub-optimal habitat') is better than the alternative, if the alternative is extinction.

The kite's tale underlines, once more, the great adaptability of some birds – which is much greater than that of butterflies, for example. Butterfly conservationists go to extreme lengths to create the perfect habitat for butterflies, but their preferences are so exact, and so poorly understood, that sometimes the butterflies still do not thrive. Birds can be the opposite. We destroy the habitat they want to live in, or exclude them from it by killing them, but in many cases they respond by finding somewhere else that isn't perfect, but will just about do.

The South Island Takahe of New Zealand, which looks like a giant moorhen, is an exemplar of this adaptability. A bird of dense lowland forests near water, it was declared extinct in 1898 after it could no longer be found in suitable areas. But in 1948 it was rediscovered in a completely different place – high mountain meadows. The habitat was not ideal, since the food was much lower in nutrients than where Takahes used to live, so they struggled to eat enough. But it was sufficient. Takahes – and Red Kites – have survived by making the best of what they have.

So please raise an arm to cheer the happy return of the Red Kite – while, of course, keeping a watchful eye on your washing line to make sure your underpants are still there.

GOLDEN EAGLE

The king of the birds is deposed

The powerful eagle has been the bird most liable to be pressed into service to add the required awe to the flags and assorted symbols of a hundred kingdoms and empires, ever since the Sumerians used it five millennia ago to grace the doorway to a royal palace.

Single-headed eagles adorn the coats of arms of Germany, Austria, Egypt and Nigeria. Double-headed versions – which add that extra air of power when a mere single head will not do – have served as symbols for Russia, its ancient enemy Serbia, and the Byzantine Empire. In the Roman army the loss of a legion's eagle was the most heinous symbol of disgrace – and *The Eagle of the Ninth*, the classic 1954 children's book by Rosemary Sutcliff which reached the silver screen in early 2011, describes a young man's quest to save the legion's honour by retrieving it. Geoffrey Chaucer summed up the eagle better than all poets who came after him, when in the fourteenth century he described 'the royal eagle' that 'with his sharp look pierces the sun'. Chaucer was referring specifically to the Golden Eagle, and when you train your binoculars on a Golden Eagle circling high in the sky, to gaze at its indignant-looking, arrogant expression, you can see what he meant.

The eagle was also high in the avian hierarchy among the Ancient Greeks and Romans, who looked at the behaviour of birds when making important decisions such as whether to fight or delay their attack – the word 'auspicious' comes from *auspicium*, the process of judging birds' flight and voice to judge the likely success or failure of ventures (and *auspicium* comes from *aves spicere*, 'to look at the birds'). In Homer's *Iliad*, written in about the eighth century BC, the Trojans ignore the bad auspice of an eagle, soaring above them, which throws a caught snake into the army after receiving a nasty bite. Bloody defeat follows at the hands of the Greeks.

The Golden Eagle is the king of the birds in nature too, capable of eating almost any bird. Even fellow birds of prey, though rarely consumed, sometimes face the indignity of having their food stolen by this raptor.

But the arrival of modern firearms brought the Golden Eagle firmly down to earth. By the end of the nineteenth century the eagle and other birds of prey were in trouble in Britain, following the systematic and highly successful extermination programme carried out by the upper classes on these birds that

dared to spoil their sport by eating their game. To protect grouse and pheasants, gamekeepers destroyed these birds in huge quantities. Early conservationists started to raise objections, but had powerful enemies against them.

Into the debate stepped Thomas Powys, fourth Lord Lilford and mid-Victorian president of the British Ornithologists' Union.

It is easy to write off milord as a harmless and rather ineffectual English aristocrat – a man of similar background to fellow Old Harrovian and birdwatcher Richard Meinertzhagen (more of whom later in the Meinertzhagen's Snowfinch essay), but without the same dark side.

Lord Lilford was an amiable soul who never seems to have done anyone any harm between the time he came into the world in 1833 and the point – weakened by repeated attacks of that traditional malady of the English nobility, gout – that he waved a languid aristocratic goodbye to it in 1896.

Lord Lilford bore his share of ill luck. A species of woodpecker named after him when he shot it in Greece turned out not to be so new after all, but the already known White-Backed Woodpecker. Likewise, a freshly discovered crane given the Latin name *Grus lilfordi* in his honour much later in life is no longer regarded as a separate species either, but a race of the Common Crane. Then, during a jaunt to Europe he travelled to the kingdom of Sardinia to fulfil his ambition of shooting an ibex, only to be told that King Victor Emmanuel had recently decided to restrict this prerogative to himself. Lord Lilford contented himself with a chamois, so presumably they never ran out of shammy cloth at his inherited country pile, Lilford Hall of Northamptonshire, to which he later retired to set up an ever more ambitious aviary.

His zeal for keeping menageries seems to have started when he was at Harrow, to have grown when he went up to Christ Church College, Oxford, and to have achieved its full flowering at Lilford Hall. There he kept lammergeiers – small vultures that can be seen in the Pyrenees amongst other places – and there was also a button-quail, a small, drab shy bird that keeps to cover and presumably sheltered under the mahogany dining table at Lilford Hall. It used to scare house guests with a loud booming call in the middle of the night.

But Lord Lilford was, despite his eccentricities, a quiet revolutionary in English ornithology – one of the key influential figures who started to turn public opinion away from shooting birds of prey and towards preserving them.

Lord Lilford's magnum opus, the three-volume *Figures of British Birds* completed in 1897 after his death, looks typical of works of the time. The birds are posed with the same aristocratic sangfroid that Lord Lilford shows in a photo of himself in his spacious library. The Reed Warbler, which in reality skulks in swamps making an odd sound like a cross between scraping and laughing, is standing to attention at the top of a reed bed as if on parade with the Northamptonshire militia, which Lord Lilford briefly joined on the outbreak of the Crimean War.

But when you start reading the book you realise how epoch-changing it is. The first volume is devoted almost entirely to birds of prey, and drips with acid observations on humanity's proclivity for killing them. For example, 'The Rough-Legged Buzzard breeds commonly in the north of Continental Europe, and is by no means an infrequent autumnal visitor to Great Britain, where it is generally destroyed very soon after its first appearance.' As for the White-Tailed Eagle, which was soon to be extirpated in Britain before its eventual reintroduction many decades later, 'as I should be very unwilling to be the cause of the further molestation or destruction of this fine species, I refrain from publishing the very little that I know concerning its nesting localities in this country.' This was a man staging a quiet rebellion.

What did Lord Lilford achieve, you may ask? Part of his success was in working behind the scenes within the British elite to get things done quietly. For example, as a member of the House of Lords he was able to use his power to give legal protection for the first time to owls. But just as importantly, by forsaking the conventions of his class, he helped to turn the public from shooters to observers of birds. Given his achievements, it is a pity that both attempts to immortalise him through bird names were so spectacularly unsuccessful. Never mind – at least there's a pub in Lancashire named in honour of his family, which is probably a more effective way to have your name remembered in England.

KESTREL

With birds on our side

Why do so many warriors bedeck themselves in bird feathers?

One obvious reason for using birds as bling is that feathers make them look bigger. The Masai are a pretty tall bunch anyway, but the ostrich plumage which they wear in a broad circle round their heads makes them look taller still. That's intimidating. It's the same principle as the bearskin caps that the soldiers outside Buckingham Palace wear, or the two long pheasant feathers often depicted projecting from the helmet of Lü Bu, the ancient Chinese general – curling far over his head and adding a good foot to his apparent height.

Bird feathers were also, historically, what the well-dressed warrior was wearing because birds were hard to catch. An eagle had to be hunted with great skill, so bearing an eagle's feathers was a symbol of prowess – or possibly of power, since a cynic might assume that Native American chiefs, who had more than the normal share of feathers, let younger and junior warriors do most of the hard work of actually catching the eagle.

Birds' association with the gods also appeals to warriors. It is better to fight 'With God on Our Side', as the Bob Dylan song puts it, armed both with moral justification and with supernatural powers bestowed by deities. Because most birds can fly, they are seen as capable – far more capable than land-based creatures – of reaching the celestial sphere. Early European religious paintings frequently depict an eagle soaring in the sky next to Christian saints. Some cultures have gone even further, and treated birds as gods themselves. To the Tlingit people of Alaska, the raven was traditionally the chief deity, and the Ancient Egyptian god Horus was depicted as a human figure with a bird of prey's head.

But birds' connection with war goes beyond warriors simply wearing their feathers. Birds are frequently invoked in the language of war, and none more so than the kestrel and other falcons.

The kestrel is a rather slim, dapper-looking bird of prey. The male has a bluish-grey head and a slightly darker downward-sloping moustache. Its sharply pointed wings, like those of all falcons, are built for speed. It used to be Britain's commonest bird of prey before a recent decline thought to be linked to a fall in the number of voles on which it feeds. But the kestrel has a particular idiosyncrasy that endears it to us – a strange liking for motorways. This allows kestrels to lend a talon to stave off the boredom of nature-loving car-bound children who can be drilled to keep eyes peeled to spot them quartering the ground in search of victims. Naturalists don't exactly know why kestrels like the grass verges beside motorways so much. One explanation is that they are not farmed, so these insecticide-free patches of land harbour more small creatures to eat, whilst another is that the cars' vibration brings earthworms, which are tasty snacks for kestrels, to the surface.

But the kestrel's great avian party trick is its amazing ability to hover. It hangs in the air, manically beating its powerful wings to keep in the same patch of sky, while it searches for prey. Other British birds can do this, but not for the same sustained length of time as the kestrel. As passengers hurtle up the M1, they leave the frantically flapping yet strangely stationary kestrel behind them. When Lockheed Martin designed a new helicopter in 2005, they fittingly called it the Kestrel in the bird's honour.

This fascination among warriors with falcons such as the kestrel has survived into modern times. In the seventeenth century Andrew Marvell used the falcon to celebrate Oliver Cromwell's victorious return from campaigning in Ireland:

> So when the falcon high
> Falls heavy from the sky,
> She, having kill'd, no more does search
> But on the next green bough to perch

Two and a half centuries later the falcon was again called into service in the muscular poetry of Julian Grenfell. Grenfell is a controversial figure nowadays because he is rare among Great War poets in failing to condemn the futility of war – and this may explain why his small but brilliant output of poetry is no longer well known. A man who described war in 1914 in a letter back home as 'like a big picnic' would very possibly have become disillusioned, like his more famous colleagues Siegfried Sassoon and Wilfred Owen, but as this brilliant young man was killed by a shell less than a year later at the age of twenty-seven, fate did not give him that option.

'Into Battle', Grenfell's best poem, evokes the warrior's heightened senses on the eve of going over the top – a state of extreme acuity that might just save his life. Searching for the best way to describe his sensation, Grenfell falls back on birds, invoking them as if they can come to the warrior's aid:

> The kestrel hovering by day,
> And the little owls that call at night,
> Bid him be swift and keen as they,
> As keen of ear, as swift of sight.

Reading these poems written so far apart in time but with such similar avian symbolism, I wonder whether it is the awesome powers of birds, inimitable by humans, that appeal to the nature lover as much as the beauty of their plumage.

Fifty-three years after serving as a symbol in the poem of an aristocratic poet, the kestrel provided a means of escape from the persecution of daily life for a working-class boy in Barry Hines' 1968 novel *A Kestrel for a Knave* – later turned into the film *Kes* by the director Ken Loach. This time, however, it was the kestrel's turn to die – murdered out of spite by the bullying elder brother of Billy, the boy who had been training the kestrel. Life's cruelties are classless, and so are the symbols we find to console ourselves over them.

SPARROWHAWK

Competing with Tibbles

Sparrowhawks – which are more likely to be found in British gardens by the eagle-eyed (or sparrowhawk-eyed even) than any other raptor – have a handy tip for marital bliss. The females of this round-winged, attractively striped, astoundingly agile bird of prey are about one-quarter bigger than the males.

Why would larger females make married life any easier for a male sparrowhawk? It sounds distinctly unpromising for them. However, it does allow males to catch different, smaller prey than females. The male contents himself with lesser birds like the chaffinch and Great Tit, which he diligently and dutifully often decapitates and plucks before passing to the mother bird if she's going to feed it to the chicks. But the female is more ambitious. She goes after starlings, blackbirds and even woodpigeons – doubling the feeding opportunities of a pair of sparrowhawks by allowing them to fill two ecological niches at the same time. The phenomenon of larger females, which is particularly common among birds of prey since their quarry is dictated by their size and power, is the avian equivalent of the rhyme of Jack Sprat and his wife. Jack Sprat, incidentally, is an old nickname for a small man, derived from the old meaning of 'jack' as 'small'. Thus the Jack Snipe – a British species of wader considerably smaller than the more widespread Common Snipe.

This division of labour makes garden life sound ideal for the sparrowhawk, by allowing it to catch virtually all birds found in British gardens. It can certainly be an efficient killer – which explains why the musket, an early type of gun, is named after an old moniker for the male sparrowhawk. But something casts a shadow over the sunny picture of perfection: it is the sinuous shape of the household cat.

Cats kill an enormous amount of garden birds – not because they are exceptionally effective, but because there is an enormous amount of cats, estimated at about eight million in the UK. We don't know exactly how many birds they catch. But in Wisconsin, where a plucky professor of ecology, Stanley Temple, received death threats from cat lovers for investigating the issue, it was estimated that up to 200 million birds were being killed in the state every year by up to two million cats. Some bird lovers argued that his findings put the cat into 'catastrophe'.

But does this affect the populations of the small birds that cats kill? This may sound a silly question, but it's not clear that they do, at least in the long term. Small birds produce large numbers of young every year, and every year the majority of these young die, while enough survive to produce the next generation. There are far more dangerous phenomena that could affect the long-term survival of a species, such as reducing its habitat, whereas the gradual attrition of young birds to a moderately but not 100 per cent effective predator such as a pet cat is probably not enough to make a difference. Although cats have been known to exterminate a whole species when introduced onto an island where flightless birds have evolved to possess absolutely no effective resistance against them, we are not talking about flightless birds here.

What is more likely, however, is that because cats catch so many small birds, other predators such as the sparrowhawk lose out. In a garden a sparrowhawk may be watching for young birds that foolishly expose themselves to danger, but it is competing with a cat. So if there were fewer than eight million cats in Britain, there would probably be more than the current 40,000 pairs of sparrowhawks.

One solution advocated by some bird protection societies in the United States is to curb cats by keeping them indoors – though it is striking that in Britain, where cats hold an especially revered place in society, calls by conservation organisations to restrain feline social life are more muted. Many cat-owners could also counter-argue that they are balancing Tibbles' bird extermination by keeping garden bird feeders in the winter. This is particularly the case in Britain, where the *Daily Star* once underlined the widely followed winter ritual of putting food out for the birds by leading with the exhortatory headline 'LARDS OUT FOR THE TITS'. You need to say that very, very slowly if you're repeating it to a friend, as I can testify from bitter experience.

The sparrowhawk can console itself that if in this earthly sphere it has to put up with rivalry from cats, in the world of Greek mythology it has a much happier lot. Hawks bear the family name *Accipiter*, after the Latin for 'to seize', but the sparrowhawk's species name, *Nisus*, is even more intriguing. Nisus, king of Megara, was turned into a bird of prey after the city was betrayed by his daughter Scylla. What happened to Scylla, you might ask? She was turned into a lark, condemned for ever to fly around in fear of her father. Revenge is a dish best served for eternity.

LITTLE OWL

Companion to a goddess

The Little Owl is one of the few identifiable bird species with a major supporting role in a Hollywood movie.

The origin of its good fortune? An even greater honour – it was a symbol of Athene, Greek goddess of wisdom, who appears in the 1981 film *Clash of the Titans*. Hence the Latin name *Athene noctua* – night-time Athene.

This 22-centimetre paradox – cute but ferocious-looking at the same time, like an extremely angry two-year-old child – was a constant companion of Athene in ancient lore. As a symbol of the goddess, it featured on coins minted almost 2,500 years ago, and can be seen in the remains of a mosaic at the excavated Roman baths at *Aquae Sulis* – the ancient name for the town of Bath in Somerset.

The mechanical owl in the film – given to Perseus by Athene to accompany him on his adventures – developed a celebrity of its own, making a cameo appearance in the 2010 remake, just as human movie stars often appear briefly in someone else's picture. But is it actually a Little Owl? The real thing is a cute chocolate-brown creature, with a breast that looks a mixture of light and dark candy. The film's owl is a little lighter, and is given the name

Bubo, in possibly the strangest ever example of Americanising a story for Hollywood. Bubo is a conveniently cute alliteration, but also the family name for a group of owls that includes a large American species, the Great Horned Owl. That explains the origin, but purists would be shocked. To start with, the Great Horned can be more than ten times the weight of Athene's diminutive companion. Perseus would not want to get into an argument with it.

The Little Owl's British existence is equally based on artifice. It is one of the handful of birds that has prospered after a deliberate introduction into this country, though after some false starts (for a discussion of the virtues and vices of introductions, see the Ruddy Duck essay). It is also one of even fewer introduced birds that look quintessentially British by merging into the muted colours of our countryside (unlike the pheasant, for example). By the 1920s it had become firmly established, after various attempts that started in Victorian

times with the notoriously eccentric Yorkshire landowner Charles Waterton (a man alleged to bite the legs of his guests underneath the dinner table).

Why was there such a fixation with establishing the Little Owl in England? It was probably its association with Athene, at a time when most educated people still felt that the Ancient Greeks represented the pinnacle of civilisation. So the Little Owl is the feathered equivalent of the pseudo-Grecian columns to be seen even in early twentieth-century buildings.

People might ask why the Little Owl is associated with Athene and wisdom in the first place. The origin lies, perhaps, in its large, rounded head, which has a strangely human appearance, accentuated by its two eyes staring in the same direction. But Little Owls are not particularly clever by bird standards. They have even been recorded falling over backwards while pulling earthworms out of the ground, though if I lift the lid on this incompetent slapstick I should counterbalance it by revealing also their ingenious habit of storing beetles by impaling them on thorns, perhaps saving them up for a day when food is scarce, or possibly because they prefer their food dried by the wind, in the same way that people dry fish and meat.

Although the Little Owl is associated with wisdom, this does compete with an older and darker British view of owls that lingers, where these birds of the night are emblems of loneliness at best, and death at worst. Thomas Gray quickly painted the sad atmosphere of his 1751 *Elegy Written in a Country Churchyard* with the introductory observation that 'The moping owl does to the moon complain', and his contemporary audience needed no more scene-setting after this symbolic shorthand (although the Little Owl is a relative rarity among owls in that it can sometimes be seen hunting by day).

The British are not the only people to regard owls with suspicion. In Sicily, for example, if a Scops Owl – small and ferocious-looking like a Little Owl, but with sticking-out ears – visits the house of a sick man, he will die three days later according to local lore.

Owls' other-worldly voices have not helped their image. The verb 'ululate' – to wail or lament loudly – comes from an Ancient Roman word for owl,

Ulula. I could cite a reference from classic Roman literature, but quoting from *Carry on Cleo*, the classic 1964 British comedy, is probably more fun. 'I hear the night owl screeching,' asserts Kenneth Williams (as Caesar) as he fakes the delirium of a dying man about to pass from this world into the next, in a comic scene where everyone is labouring under the misapprehension that he is dying. Actually, the sound is only his rather shrill wife – Joan Sims playing Calpurnia. The henpecked Caesar character of the film knows this, but he has used the owl to exact a little revenge by way of insult. The joke worked because the scriptwriters of suburban Pinewood Studios had a couple of thousand years of rich owl symbolism with which to play.

SNOWY OWL

When it looks just like the real thing…

Strolling round the National Gallery in London is a depressing affair for any bird lover.

Most of the birds in Old Masters are so badly drawn that it is impossible to tell what they are. Strange white fowls in the background of Renaissance paintings could easily be either doves or egrets – and given that the first bird has short legs and a short bill and lives on land, whilst the second is the opposite on all three counts, this is a pretty damning indictment. Let me single out one particularly frightful painting, by the seventeenth-century Dutch artist Melchior d'Hondecoeter, for especial, well-deserved opprobrium: his 1668 *Birds, Butterflies and a Frog among Plants and Fungi*. A bird in the foreground could be a woodpecker, finch or owl, since it has the features of all three. One almost marvels at the artist's skill in cramming so many birds that are so spectacularly different into a depiction of just one.

There are some well-drawn birds in some pictures, but these are unfortunately all dead – pheasants and other game hanging in the larder. Artists were not able to capture the likeness of living, breathing fowl. Some accepted this weakness by choosing not to portray them at all – try looking for a bird in the sky or on a lake in any of the wide vistas in which the seventeenth-century French master Claude Lorrain specialised. Their absence speaks volumes.

But in the early 1800s, this suddenly changed. A small group of artists emerged who had the knack for painting live birds that actually looked like live birds – and bird illustration entered a new era. One of this small group was Edward Lear, who by a strange quirk of fate is best remembered not for his bird portraits but for a children's poem with a bird in the title, 'The Owl and the Pussy-cat'. It is a great pity that he gave up drawing bird portraits while still in his youth, because they are better than anything that had come before. His picture of a pair of Snowy Owls, drawn for the avian entrepreneur John Gould's *Birds of Europe*, is superb. It not only captures the plumage details with scrupulous accuracy, but also conveys the eternally alert quality of Snowy Owls in the wild – something that is hard to do in paintings. When you first see these great white birds from a distance they look like sheep, rocks, or large white plastic bags that have blown onto the wildest parts of the moor because of the fecklessness of some past picnicker. But after you train your binoculars on them, you can see that however relaxed they seem, they are always keeping an eye on things – just as Lear's painting conveys.

Lear was one of those character types we've all met: someone who is immensely talented in many different ways but still makes a mess of his or her life. Finding it hard to relate to adults, he was continually falling out with his colleagues. His list of enemies acquired over a long life of sulking and counter-sulking included John Gould – though to be fair, Gould, the kind of man who would always make sure he ended up with the contents of the Christmas cracker after pulling it with a friend, was an easy character to fall out with (to read about this tough entrepreneur's weakness for hummingbirds, see the Bee Hummingbird essay). Lear eventually escaped permanently to Italy with his loyal chef Giorgis, who by all accounts was a much better friend than cook. But despite or perhaps because of Lear's inability to deal with adults, he was great at comprehending the essence of both children and birds. He possessed an unfathomable ability to understand what children find funny and an equally uncanny knack for understanding how birds move and interact with their environment.

But why are pre-Lear bird paintings (and, to be fair, a good many post-Lear bird paintings) so bad? The most obvious answer is that they were usually drawn from dead specimens, which helps to explain the wooden quality of so many. Sometimes the colours are exactly right – and so they should be, if the bird being painted cannot fly away – but a bird's behaviour cannot be captured by a museum specimen. Perhaps this explains the anomaly of why early paintings of the goldfinch, a common cage bird that was alive but could not fly away, are often rather good.

Another common fault in early bird illustration was the tendency to depict birds in rather a noble pose. Many old bird paintings show them bolt upright, like children told to sit straight for their portraits. This partly reflects the posture usually chosen for dead birds that were stuffed and mounted – it looked better to have a dead goose standing to attention in one's drawing room than one with its neck down as if doing something natural but pedestrian such as chomping the grass.

What surprises is the staying power of badly drawn birds. Reproductions of old pictures of lordly-looking wildfowl and warblers regularly appear on dinner-party placemats even now. Perhaps people prefer the fiction of these aristocratic-seeming, confident-looking birds, to the more furtive features of the average bird in the wild, its eyes constantly darting around looking for predators as it feeds.

Lear, who painted live birds whenever possible, has another, totally unexpected memorial to his name. A book he illustrated on the parrot family contained what he was told was a Hyacinth Macaw. But the macaw, drawn with admirable accuracy by the great artist, can be confirmed retrospectively as a separate species, which is now known forever as Lear's Macaw in his honour. Accurate bird illustration, whether of unmistakable birds like the Snowy Owl or trickily similar macaws, can ensure immortality – which is something they don't teach you in art school.

STELLER'S SEA EAGLE

Don't believe everything you hear

The Steller's Sea Eagle – a huge bird of prey with an enormous orange bill – has become one of the world's most unlikely tourist attractions, rivalling the mathematical genius of the sixteen-sided house in Exmouth, the poetry of the Halona Blowhole in Hawaii, and the sheer culinary utility of the globe's largest rolling pin in Wodonga, Victoria.

The easiest way to see the eagle in the wild – and believe it or not, this is indeed the easiest way – is to get on a boat in the middle of a cold winter's night at the fishing port of Rausu, on Japan's northernmost island of Hokkaido. You then huddle round a stove inside the small vessel for about an hour and a half – it's best to keep inside as much as possible, because the temperature can plunge to –30°C outside – while the boat ploughs through choppy seas to reach a patchwork of ice floes somewhere between Japan and Siberia. (Another tip: don't under any circumstances take off either of the two pairs of specially ordered Arctic-weather gloves you're wearing.) Then a taciturn man with an understandably slightly glowering expression starts throwing out a strange concoction that rejoices in the inappropriately jolly name of 'chum' – oily fish ground up and scattered over the ice.

At this point the eagles appear in large groups – unusually large groups for birds of prey, since you can see tens at a time. They chomp contentedly on the fish, the tourists take photos, and then the boat begins the return journey to Rausu, disembarking you in time for a hearty breakfast of raw fish and swollen crab testicles – a local delicacy. All in all, it adds up to a splendid morning.

This pilgrimage to see the eagle has become a popular diversion for adventurous travellers in recent years – a favourite not just with birdwatchers, but also with pure aficionados of what is sometimes called Extreme Tourism. Most of the voyagers are a mixture of the two: deprived of real adventure and a test of physical endurance in their day jobs, birdwatchers from rich countries seek it outside. More generally, this touch of hardship is one of the appeals of international birding, which adds a dangerous edge that a visit to a local nature reserve, ending in a leisurely exit via the gift shop, does not offer back at home.

It is ironic that a bird which seems the very embodiment of wildness has benefited so much from human interaction. The commercial fishing fleets that depart from Rausu get a good catch, the eagles take their share of it by following the boats, and the eagle chums' chum from the tourist vessel provides a little extra. The bird's global numbers have increased as a result to about 6,000, of which about a third winter in Japanese territory after breeding in Siberia. It is not quite the largest eagle in the world, but it is the heaviest, and is that really a surprise with all that fish inside it?

Quite apart from its modern incarnation as a tourist attraction, the Steller's Sea Eagle, named after the German naturalist who furnished the first proper scientific description of the bird in 1811, has very possibly been exciting our imaginations for a long time before that. The Steller's Sea Eagle is one of the prime candidates to have provided the inspiration for the roc – the mythical huge bird of prey which is supposed to have had the power to carry off and eat elephants, and to have destroyed Sinbad the Sailor's ship in the fairy tale. The roc is often depicted as white – possibly a reference to the large white flashes on the wings of the Steller's Sea Eagle – with a huge hooked beak which is also reminiscent of the eagle's.

Other possible contenders for the roc include the gigantic Haast's Eagle of New Zealand – bigger than any bird of prey that still exists – which became extinct in about 1400 after its main prey, the giant, wingless moa, died out and there was nothing large enough left to satisfy its enormous appetite.

But my favourite explanation for the roc, because it is unparalleled in pure ingenuity, is that it was inspired by the ostrich. A key feature of the roc legends is the huge size of its egg, and hence of its young – indeed Sinbad's ship was attacked precisely because the crew feasted on one of these Big Mac SuperSize objects. Could the sailors who spun tales about the roc have been confused by the ostrich? Despite being the world's largest extant bird, its flightlessness and downy feathering give it the appearance of a huge baby, whose fiercely protective mum would be pretty big indeed.

What is the antithesis of the Steller's Sea Eagle? Probably the Jackass Penguin, another recent tourist attraction that since the 1980s has shared Boulders Beach near Cape Town with the tourists. You can stroll to within a metre's distance of it, while slurping your ice cream – and you don't have to take off two pairs of gloves to eat your treat. But for those many birdwatchers who like their hobby precisely because it takes them to what Shelley called 'all waste and solitary places' that are in the middle of nowhere, where would the fun be in that?

WATERBIRDS

GOLDENEYE

The Bond bird

Does this dumpy duck have a glamorous Bond film named after it?

The role in cinematic history of the goldeneye is a point of much debate among dinner-jacketed international spies and donkey-jacketed birdwatchers alike. We know the film was named after the Jamaican estate of Ian Fleming, the black sheep of a distinguished banking family who redeemed himself in a highly unusual way by writing the successful Bond books. But why did the writer call his estate Goldeneye? Was it after this strangely top-heavy duck, with a large green (for a male) or brown (for a female) head with a bump at the apex, as if it has knocked its cranium on one of the man-made nest boxes which it favours?

Fleming himself gave various reasons, including his involvement as a real-life intelligence agent in Operation Goldeneye, a plan for the defence of Gibraltar if Spain joined the enemy Axis powers during World War Two. But Fleming was also a birdwatcher. Surely this must have played some part?

An even more resounding bombshell is that James Bond himself was named after an expert in that most uncool of male pursuits: birdwatching. The original James Bond was a professional ornithologist and author of *Birds of the West Indies* – a tome kept by Fleming at Goldeneye. Explaining his choice, Fleming said: 'I wanted the simplest, dullest, plainest-sounding name I could find.' Suddenly the intriguing scene where the Roger Moore version of Bond identifies a *Lepidoptera* specimen that M is studying seems strangely explicable. 'How does Bond know that?' I used to wonder as a child. The connection between the real and the fictional is finally made in the 2002 Bond film *Die Another Day*, where the fictional James Bond is studying the real James Bond's *Birds of the West Indies* in a scene that takes place in Havana. It's essential reading for all successful spies, of course.

In Britain, many goldeneyes have adopted the sort of distinctly metropolitan existence that James Bond himself enjoys. London is among the best places to see it in Britain, with the goldeneye one of the star turns of a trip to its reservoirs. If you're a Londoner, you don't need to worry about loading up your car with telescopes and other unwieldy gear, before driving off to some distant and obscure place to find it. You can see it just by hopping on a London bus.

But has the bird become just a little too domestic?

For most of British history it has been a purely winter visitor to our cold and windswept lakes and seashores. When the poet George Crabbe was groping for ways to describe the alienated, isolated wilderness that the evil social misfit Peter Grimes found himself cast into, as he lived a hermit's existence before going mad and dying, he wrote in 1810 of 'the clanging golden-eye' of the Suffolk marshland which Grimes inhabited. This creature of wild places with its rather loud and at times unnerving call was not a bird associated by poets with happy domesticity, like the robin. But these days, the preference of Britain's goldeneyes for manufactured nest boxes is the ultimate sign that they have been conquered by humanity.

To many people, the goldeneye's tameness is a good thing. Does it matter if it returns home to man-made nest boxes after a hard day's feeding in man-made reservoirs? This helps it to thrive, and makes it easier to find, to boot.

But other people argue that when birds become too tame, it takes away much of the pleasure of our appreciation of them. To them, birds are fun because they live in a parallel universe, where they make their own rules and lead their own lives without any reference to us. Goldeneyes, by contrast, inhabit a kind of glorified zoo. Where's the fun in that? And what would be the next step? Setting up a dating service for goldeneyes? You may scoff at such a ridiculous idea, but this is precisely what many birdwatchers proposed for the companionless female Snowy Owls that in the 1970s started summering in the Shetlands every year. They wanted to introduce the lonely ladies to a male brought by plane to the islands, and were only seen off after a lengthy debate over the ethics of interfering with nature's course.

The most important consideration, of course, is what the fictional James Bond would think of the goldeneye's new habits. He might well disown the bird with which he is most closely associated. For starters, the goldeneye is what even its own mother would call 'interesting-looking', rather than handsome like the espionage hero. Its Latin name *Bucephala clangula*, 'clanging horse's head', gives a good idea of the strange shape and sound of this mid-sized duck. Unlike the English spy, it is no Adonis. James Bond would also be reluctant to breed in a nest box, at least judging by the films I've seen. It is not a romantic setting for lovemaking – apart from anything else, there's a distinct lack of space, although it does often afford a picturesque lakeside view.

But the most unglamorous fact of all about the goldeneye is that one of the greatest global risks to its conservation is the chance of a mass infection caused by eating contaminated sewage. James Bond dines in some pretty fancy restaurants, and when he takes a dip it is usually in some billionaire's pool. He does not face hygiene problems in any of the films that I remember. On the other hand, goldeneyes do not normally have to escape the jaws of pool-bound sharks.

RUDDY DUCK

Oversexed and over here

This splendid-looking duck, with a rufous plumage and long and strangely upright tail, encapsulates one of the great dilemmas of modern conservation – not 'When will the killing end?', but 'When should the killing begin?'.

When the British government decided in 2005 to exterminate the species in Britain, many nature lovers were appalled – and were even more shocked that the Royal Society for the Protection of Birds (RSPB) had decided to co-operate. Why on earth was a conservation society killing things?

The extirpation plans spawned an underground resistance movement in support of the ducks. Fearful that the government might trawl the Internet looking for references to the Ruddy Duck, fans reporting sightings of the bird on the Net abbreviated them to 'RDs' or other such codes. An avian cyberwar erupted.

But the RSPB had sound logic for its support of the cull. Ruddy Ducks are actually not British at all but American. Imported to grace wildfowl collections, they took to the local habitat like ducks to water after escaping into the wild in the 1950s. By 2000 there were 6,000 wild Ruddy Ducks in the UK, and it's at around this time that they started looking farther afield. Soon, many were wooing the local señoritas (or señors) in Spain: White-Headed Ducks, a separate but similar species that is globally endangered but had, through Herculean conservation efforts, increased its Spanish population from twenty-two to 2,500 birds. The Ruddy Ducks threatened to reduce White-Headed Duck numbers by hybridising with them to produce offspring that were neither fish nor fowl nor White-Headed Duck. They were, in short, a Ruddy nuisance, making it impossible to duck the issue.

British cynics saw a political motive behind the British government's move: a desire not to get on the wrong side of Spain, which was already miffed with Britain over its refusal to hand over Gibraltar (though Spain had been disgruntled about this for 300 years). The European Commission had also supported the extermination – provoking accusations of interference from Brussels. Opponents of the cull argued that hybridisation with White-Headed Ducks was rare. Animal Aid, a campaign organisation, raised the historical ghost of Nazism by saying that Ruddy Ducks had been 'sentenced to death in the name of blood purity'.

But this is a case of stretching analogies too far. Ruddy Ducks are a different species from White-Headed Ducks. Quite apart from this point, introduced species have possibly caused more bird extinctions than anything else in the modern world – though more often it has been mammals such as cats, rats and dogs, rather than new birds, that did the damage. Introduced bird species often thrive, freed from whatever predator has evolved to keep their numbers down in their natural homeland. The result is that they usually displace something else. The population of the Ring-Necked Parakeet – another species introduced to Britain, this time from Africa and Asia – has shown an astonishing rise, growing by 700 per cent since 1995. Some conservationists have blamed it for a fall in Lesser Spotted Woodpecker numbers, arguing that the two species compete for suitable nesting sites. No one is talking about exterminating Ring-Necked Parakeets in Britain yet, but they may well do so in the future.

But are there exceptions to the rule that 'introduced birds' are bad?

Occasionally, yes. Britain has far more introduced species than most people realise, and many have established stable populations that have not run out of control. One example is the splendid chocolate-brown Little Owl (for more on this companion to a goddess, see the Little Owl essay). Most of all, if you have grown to love the Mandarin Duck, you will be pleased to hear that, although it was introduced from its original homeland of Asia, conservationists are glad it is here and thriving. Laypeople's opinions on the duck differ. Some dislike the contrast between the unusually showy multicoloured male and the unusually drab female – perhaps the greatest gulf in attractiveness between the sexes in the whole of the bird kingdom. Whereas many male birds have crowns or manes, the male Mandarin has both – giving it a rather top-heavy appearance. The Mandarin's elaborate headgear is commemorated in its rather snide and potentially slanderous Latin name, *Aix galericulata*, which means 'wigged duck'. The male Mandarin may not mind, however, because it appears to have a sophisticated sense of humour. There is a famous photo of it displaying while sitting on the head of Lord Grey, birdwatcher and Foreign Secretary in World War One – a Mandarin exerting its authority over the former overlord of the Foreign Office's mandarins.

But forgetting for one moment this battle of the sexes over the Mandarin, its numbers, at 7,000 and rising, usefully supplement a declining population in its Asian homeland. This makes the decision to persecute one duck and preserve another perfectly logical.

Don't tell this to any British male Mandarin Ducks though – they're big-headed enough already.

SNIPE

Never fly in a straight line

Veteran fighter pilots in World War Two used to emphasise one particular trick again and again to young recruits inexperienced in the art of killing and surviving in the air: never fly in a straight line for more a few seconds, or it will be the last time that you fly.

Pilots well versed in country sports would already have learned this from the zigzag flight of the snipe (more properly known as the Common Snipe), the streaked brown wading bird which has infuriated hunters ever since firearms were invented. Anyone can shoot a pheasant, hunters say, but to bag a snipe requires consummate skill.

The British army had already acknowledged the capability of this bird in the nineteenth century, by introducing what it called 'snipers'– crack shots taught to target, follow and shoot commanding officers, reconnaissance parties and other key personnel from a great distance. This was a borrowed word – it had been invented in the previous century as a term of praise for hunters good enough to kill snipe. Later still, it acquired its civilian meaning: a sniper will snipe at your achievements – making malicious and underhand remarks about whether you really are as good a shot as you claim, perhaps. A 'snipe hunt' is a practical joke, since both often lead the protagonist into wild goose chases, to use another bird metaphor (for another example of words and names with unexpected avian origins, read about Mavis in the Song Thrush essay).

These mysterious shorebirds, with enormously long bills roughly three times the length of their head – the longest, proportionate to their bodies, of any British bird – are sometimes hard for birdwatchers to catch in their sights too. Birders drive down to selected spots on the British coast to see large concentrations of knot, dunlin and sanderling, other wading birds that gather in large flocks for safety and fill the air with their trilling. But the snipe is usually creeping about on its own in a small pool somewhere to the side, like a solitary child at a party who finds it difficult to mix with the others.

Just like humans who keep themselves to themselves, the snipe had a mysterious and often sinister reputation in earlier times. In many parts of Europe the snipe has been seen as a bringer of rain and bad weather. The Nunamiut Eskimos called the bird 'weather maker' for the same reason.

We can all bear a little rain, but sometimes snipes have had darker associations than mere bringers of bad weather. In medieval France female snipe were thought to be incarnations of the devil's wife.

If the snipe has a reputation for being hard to catch, the Eurasian Woodcock does not, despite also having a bumpy flight. Its Achilles heel is its innate conservatism. When flying through woods it will often follow the same route day after day, which allows snares to be set to catch it. Its refusal to clamber over fallen branches that were not there before also makes it easy to set traps for woodcocks on the ground. As a result, the surname Woodcock was originally a term of insult for a stupid person.

The closely related American Woodcock has another feature much appreciated by mediocre hunters – it is renowned for being the slowest-flying bird in the world, with a stately cruising speed of 5 mph – about the speed of a brisk human walk. This may help to explain why the only record so far in Europe is of a bird found in France after it was bagged by hunters.

John Milton refers to the Eurasian Woodcock's stupidity in *Colasterion*, one of his many polemical essays against someone who had dared criticise him. Coincidentally, he later married Katherine Woodcock. History does not record whether Milton made use of her maiden name's unfortunate associations in any of their quarrelsome moments. But knowing the poet's notorious waspishness, it seems likely that she was indeed the subject of such sniping.

JAPANESE CRESTED IBIS

A Chinese success story

The name of this bird, ill-fated though not (yet) out for the count, is horrifically out of date.

You might assume that the Japanese Crested Ibis, with the scientific name *Nipponia nippon* after the old Japanese name for Japan, is a Japanese bird, commonly seen in the country's ponds, marshes and wet rice paddies, using its long, curved beak and rather short, stubby legs to pick on fish, frogs and newts.

This is now tragically wrong, although this rather strange-looking bird's equally strange and chequered history suggests that it may yet survive in an unexpected way.

In earlier times, the Japanese Crested Ibis was a bird common enough to be woven into the fabric of everyday life. It appeared frequently in Japanese literature and art, and must have been a fun but difficult bird to illustrate, with its area of pink bare skin at the front of the face, fluffy grey plumes on the head that make it look like an ageing punk who has been surprised by unexpected news and then received an electric shock, wonky long curved bill, and a body which, in different lights, looked grey or pink or white. The

Impressionists, who liked to paint the same scene at different times of day because the prevailing light made it look so different, would have loved this bird had they known about it – their enthusiasm sharpened by the knowledge that Impressionism draws much of its inspiration from Japanese painting. The Japanese responded to the uncertain colour of the ibis – neither shocking pink nor snowflake white – by dismissing men who were wishy-washy and indecisive as being 'like an ibis'. The bird was even sufficiently embedded in popular culture to have a colour named after it: a certain kind of faint red hue was known as *Toki Iro* – the Japanese for ibis-coloured – in much the same way as we use the familiar rose to describe a similar shade. One could even spot it in Tokyo Bay once upon a time, in much the same way that we might, even now, commonly see herons on a weekend London walk.

But the all too prevalent twin evils of hunting and habitat loss had reduced the bird to tiny numbers in Japan by the beginning of the twentieth century. It was thought extinct in the country in 1920, only to be rediscovered on the small and even now relatively inaccessible Sado Island, plus a neighbouring bit of the mainland, twelve years later.

By 1981 its population had dwindled to five wild birds on Sado, which were taken into the custody of the Japanese government for their own protection. Thirty years of tragicomic attempts to re-establish them have followed. Every twist and turn of the ibises' fate has been followed faithfully in the Japanese papers. Their love lives have been pored over with an assiduity usually reserved for media stars, because the ibises have now become media stars in their own right. This glare of publicity, which is so often destructive for human celebrities, has protected them – allowing numbers to rise slightly.

To achieve success, the authorities must overcome two problems. They must breed the ibises to increase numbers, and they must then release them into the wild. Both aims have met with only mixed success, despite some ingenious schemes involving Heath Robinson contraptions. Presently the ibises are kept in a huge 4,000-square-metre cage – but in 2010 martens (animals similar to weasels) got through the gaps and killed nine of the birds. Staff afterwards managed to count no fewer than 265 gaps that were large enough for a marten to get through – showing the futile zealousness in reporting the exact size, scope and anatomy of a disaster, after the damage had been done, that we see time and again in detailed government reports all over the world.

To protect the birds from predators, during night-time hours staff listen to the birds in an outbuilding some distance away through microphones, ready to rush round should the ibises' familiar croaking sounds turn into high-pitched screams. It is difficult to see how one might react effectively and with haste to a tiny marten running around the undergrowth in a huge cage, popping its head up at only occasional moments, but the ibises can at least reassure and possibly flatter themselves over the superhuman effort that is now going into protecting them – cages, bugging, and these days even an electric fence.

Fortunately for the ibis, as a species it does not have to rely merely on the well-meaning antics of the Japanese government. The ibis used to live in China too, but was thought extinct there until a colossal three-year search through the country by the Zoological Research Institute of the Chinese Academy of Sciences found seven birds in 1981. The quest continued, until by 1989 forty-six birds had been located. Rigorous conservation measures were begun, pushing up the population to about 500 birds. For once, petty nationalism (for it was this that spurred the Chinese into action in the first place) has produced a wonderful benefit. *Nipponia nippon* is thriving in the Middle Kingdom, bestowing on China a small victory in its centuries-old jostling with its traditional rival. The ongoing propaganda victory is likely to ensure the birds' continuation too. Two cheers for nationalism!

MIYAKO KINGFISHER

The bird that wasn't there

The Miyako Kingfisher is one of the oddities of the annals of ornithology for one simple reason: we don't really know if it ever existed.

My very old field guide to Japanese birds has the eccentric and rather enervating habit of including species that are no longer in the field. Its description of the Miyako Kingfisher is impressively comprehensive: 'Somewhat smaller than Ruddy Kingfisher and browner in colour with blue-green stripe extending from underneath the eye to the blue-green back, scapulars, and tail. Rump and upper tail coverts cobalt blue. Dark red legs.' Then, and only then, does it reach the killer punchline: 'Status: Extinct.' It is one of a large number of birds only known from one specimen – in this case found on Miyako Island in the Ryukyu Island chain, a long way from anywhere, back in 1887.

It may have been found in 1887, but it wasn't identified until 1919, when a Japanese ornithologist, Nagamichi Kuroda, was rifling through the Tokyo University collection of animal and bird skins. The label was decidedly vague: it just gave the name of the person who procured it, the date (5 February) but not the year, and then the name of an out-of-the-way island group – Yaeyama. On contacting the finder through an intermediary to ascertain a few more details, Kuroda received the reply: 'Miyako Island, 1887'. Miyako Island is close to the Yaeyama chain, though not actually in it.

You could say, 'Well why does that matter, if a new bird is both sketchily and inaccurately labelled? If it doesn't look like any known species, then it must be a new one.' To turn an old proverb on its head, 'If it doesn't look like a duck and it doesn't quack like a duck, then it isn't a duck.'

But here is where the problems start to multiply. The central problem is that it does look very like the Guam Kingfisher – a still-living race of the Micronesian Kingfisher – but not exactly. Its long, thick beak, typical for a kingfisher, is white, but the Guam Kingfisher's is very dark – though the Miyako Kingfisher's beak may just be white because the colour has faded after so many years. The key difference is that the Miyako Kingfisher's legs are dark red, as my guidebook helpfully informed me on the off chance that I should rediscover an extinct bird on an obscure island chain. By contrast, the Guam Kingfisher's legs are brown.

An unlucky added complication is that the man who collected the specimen, Antei Tashiro, had actually been to Guam. Given the vagueness of the label, the suspicion is that he wasn't sure where he got the bird in the first place and was blessed with even less certitude thirty-two years later. Another question is why Tashiro didn't himself declare the Miyako Kingfisher as a new species after first finding it. Was that because he didn't think it was, or because he simply didn't think about it one way or the other?

Other theories are that the bird was found on Miyako Island but had simply strayed from Guam, or was a Guam Kingfisher that had been taken to Miyako and kept as a tame animal, but then escaped.

Some of the sceptics' theories should themselves raise the hackles of a sceptic. Many people across the world retain ducks as tame ornamental species, but few keep kingfishers. Moreover, if we believe the sceptics we have to agree to the strange double coincidence that the Guam Kingfisher which made the unusual trip to Miyako was an unusual specimen with differently coloured legs. This violates the principle of Occam's razor – that we should generally accept the theory with the lowest necessary number of assumptions. In this case we are forced to make two.

This is far from the only story of a bird whose very historical existence is disputed. Lerch's Sapphire, a hummingbird species found in South America in the nineteenth century and known only from a single specimen too, is now thought to be a mere hybrid (though naturalists can't decide which birds it is a hybrid of). There is also Townsend's Dickcissel, an 1833 specimen once believed to be a new species but now considered an aberrant form of the ordinary Dickcissel of the Americas, which simply lacked the bird's usual yellow colouring. Last (but only for the moment) and certainly not least is Cox's Sandpiper, an Australian wader described as new to science amid great excitement in the birdwatching world in 1982 and then re-described as not remotely new to science, but simply a hybrid of Pectoral and Curlew Sandpipers, in 1996.

This litany of wishful thinking testifies to the innate human desire throughout history to discover new things, which sometimes causes caution to be cast to the winds. On a more prosaic level, almost any birdwatcher has a guilty little tick somewhere on their life list of birds seen – a bird identified as something very rare in a moment of excitement, but which, on later reflection, they know in their heart of hearts was something much commoner which they have seen a hundred times. So most birdwatchers have their own metaphorical Miyako Kingfishers.

There is, of course, the remote chance that another Miyako Kingfisher will turn up again – not on the small and densely populated Miyako Island, where it would have been noticed, but perhaps on an even more remote isle where it's been overlooked. To scoffing sceptics I present the case of the Large-Billed Reed Warbler, previously known only from one specimen in India in 1867, but then re-found in Thailand – a very different part of the reeds, as a Reed Warbler would say – in 2006. If I find it, I promise not to wait thirty-two years before telling anyone.

GREY HERON

Hamlet's riddle

It is one of the most unmistakable of all British birds – a 3-foot high creature that stands immobile in the shallows quietly waiting for fish. I'd wager that of all British birds, it's the one that can be identified from the greatest distance, courtesy of its characteristic flight profile. Huge bowed wings provide the power, while long legs trail behind, like a plane whose pilot has forgotten to pull up its undercarriage.

Ironically, our very familiarity with the Grey Heron has embroiled us in endless etymological confusion over the centuries. The slow grinding process of eliminating the plethora of names granted to this bird is a textbook example of the gradual victory by scholars and their servants – headmasters and headmistresses – in eliminating the rich variety of local words and spellings for animals and objects in favour of one common standard. It is an inexorable form of lexical musical chairs that leaves only one word left with a place.

The most famous confusion is in Britain's most famous literary masterpiece, Shakespeare's *Hamlet*:

> I am but mad north-north-west: when the wind is southerly I know
> a hawk from a handsaw.

So says the Danish prince to Rosencrantz and Guildenstern – explaining that although he is acting like a madman, he can return to sanity as quickly as the wind changes. A handsaw was a variant of the more common 'harnser', an old word for the Grey Heron, rather than something used for cutting wood. Comparing hawk with heron, which is no easy bird even for a bird of prey to kill, seemed natural at a time when the heron was seen as the perfect quarry that would allow keepers of tame hawks to show off their birds' skill.

Another old name for the Grey Heron is 'crane', which has caused no end of trouble since what we call the crane today is another family of bird altogether. This fact has even prompted arch-sceptics to suggest that cranes never bred in Britain until their recent recolonisation in East Anglia – they were simply herons. There is a faint possibility that they are right. However, references in the lists of fowl served at medieval and Tudor banquets – one of the key sources of evidence of what birds were in Britain in early times – sometimes refer to herons and cranes. This suggests that both were in the country until the 1500s, when the crane was extirpated in Britain for a few hundred years. During the crane's absence, English became heavily standardised, thumped into a useful but drab sameness in the centuries that followed Samuel Johnson's weighty *Dictionary of the English Language*, published in 1755. Old, affectionate names for the heron that reflected its familiarity and rather human dimensions (by bird standards), such as Julie-the-bogs, Old Nog and Frank, gradually disappeared.

But the process of standardising the English language was in fact *begun* rather than ended by Johnson, and is still going on within our lifetimes. Taking birds as an example, even during my 1980s childhood I knew experienced birders who refused to kow-tow to the suggestion of the RSPB and most field guides that we should call a Little Grebe a Little Grebe and a dunnock a dunnock. Linguistic rebels referred to the former as a dabchick – which the RSPB disliked because it wanted the bird's family name to be reflected in its title. Some diehards even referred to the dunnock as the Hedge Sparrow – the name it was most commonly known by for centuries because people mistakenly thought it was indeed a sparrow. Many of these old names were far more charming than the modern equivalents.

Those of us in what historians will doubtless remember as the *Bagpuss* Generation – people who as children watched the cult 1970s television programme about a cloth cat – may remember Professor Yaffle, the bookend carved into the shape of a bird of uncertain species. Yaffle is in fact an old onomatopoeic name for the bird that now goes by the rather unimaginative moniker of the Green Woodpecker. The word is an imitation of its mocking laugh as it flies from tree to tree, evading visual capture by eager birdwatchers. It was a name with a story – a name that would prompt a curious schoolchild to ask, 'Why on earth is it called that?' But the conformists have won, both in ornithology and in most other areas of life, and one rarely hears a Green Woodpecker called a Yaffle anymore. Will it go even further? Having almost eliminated Hedge Sparrow from the English language, purists would like us to call the dunnock the Hedge Accentor, since it is part of the Accentor family (from the Latin for 'chorister', since many accentors including the dunnock have fine warbles). They are currently losing this particular war, but for how long?

Conformists have certainly won in matters gastronomic too. TV chefs may have a strong liking for exotic Oriental fare but most would baulk at a heron. However, in medieval times the heron was regarded as tasty enough to serve on a dish for aristocratic gourmets. Curiously, Britons' palates seem to have changed, since by the twentieth century anyone who ate it regarded the bird as almost unbearably fishy – though I doubt the heron sheds any tears over that.

GREAT CRESTED GREBE

From watching birds to watching people

The Great Crested Grebe is one of nature's show-offs, but like many show-offs, it eventually paid the price for its big-headedness.

In breeding plumage males and females look equally attractive, with black crowns and rufous manes below them that darken once more into black at the bottom. This unusual parity between the sexes is explained by the fact that they play an equal part in courting each other, unlike the normal state of affairs in the bird world, where the male does the displaying and the female plays the part of the bashful bride-to-be.

The Great Crested Grebe's Weed Ceremony is one of the most celebrated of all avian courtship rituals – and is amazingly PC since the male does not take the lead, though it probably does not meet today's strict health and safety requirements given the involvement of grubby pondweed. Male and female swim slowly away from each other, making piping sounds, before diving. Both resurface with a bunch of vegetation in their bills, rapidly approach each other, and bring the ceremony to a climax by rising in the water, furiously treading water like cartoon characters who have fallen off a cliff, and rocking their heads rapidly from side to side. Who says true romance is dead?

But it was the plumes of its crest that very nearly put paid to the Great Crested Grebe, when Victorian ladies started thinking they would look just as good on their own heads as on those of the birds. By 1860 there were just forty-two pairs left in Britain.

If every action produces an equal and opposite reaction, the persecution of the grebes presented a perfect example. In 1889 a group of well-to-do women gathered in Didsbury, a Manchester suburb, in an effort to stop the use of birds in fashion. Motivated largely by the persecutions in particular of the Great Crested Grebe and the kittiwake, a gull with an onomatopoeic name, they formed what was first called The Plumage League and later became the Royal Society for the Protection of Birds – today Europe's biggest nature conservation organisation, with a membership of over a million. The Great

Crested has prospered too, though a little less spectacularly, with a present-day population of just under 10,000 breeding pairs.

The Great Crested Grebe Inquiry of 1931 was co-organised by the anthropologist and ornithologist (and Old Harrovian, like an enormous number of distinguished ornithologists of his day) Tom Harrisson to see how the birds' fortunes had revived. This helped inspire him to co-found Mass Observation, the influential but controversial study into social attitudes. Mass Observation was initially conceived, in reaction against the national newspapers' criticism of Edward VIII's desire to marry the divorcée Wallis Simpson during the Abdication Crisis, by Harrisson and other like-minded academics who wanted to dig beneath media hysteria to survey the true mood of the public towards Edward VIII and Mrs Simpson. But it became a crucial way for the government to monitor the morale of ordinary Britons during the Blitz of World War Two. Harrisson managed to enlist the enthusiasm of several leading birdwatchers for MO, including Max Nicholson, James Fisher and Richard Fitter – suggesting that many birdwatchers enjoy being people-watchers to the same degree.

The style of Mass Observation was largely shaped by the study of birds. Volunteers were asked not only to keep their own diaries, but also to listen out for interesting and revealing snippets of conversation heard on the street that reflected everyday behaviour and thought – much as a birdwatcher might watch grebes and note down any interesting actions. Harrisson felt that what worked for birds worked for people too. As he put it: 'You don't ask a bird questions. You don't try to interview it, do you?'

Nowadays, birdwatching has gone way beyond spying with binoculars: the nationally known ospreys that breed at an RSPB site in Loch Garten can be seen live on a webcam, as can the goshawks and hobbies that nest in the New Forest. Digital snooping on such birds keeps them safe from egg collectors and vandals by putting them on public view all the time.

But Mass Observation has also been criticised as marking the beginning of Britain's surveillance society, since it involved monitoring people without their knowledge or consent. We moved from spying on grebes to spying on our own citizens. Is the same treatment OK when practised on humans? Intelligence agencies can fairly argue that a certain amount of watchfulness is necessary in the interests of national security.

But regardless of the moral arguments for and against surveillance, the decision by the founders of Mass Observation to treat *Homo sapiens* just like any other species is deeply significant, because it marks scientists' changed view of humanity in the post-Darwin age. These days our species is regarded by many scientists as just another among thousands, rather than a special case marked out by God as different from other creatures.

RED-CROWNED CRANE

Teaching birds to be wild

It was schoolchildren who saved the Red-Crowned Crane.

This tall, stately, long-necked bird of serpentine beauty has a magnificent mating dance that begins rather shyly but ends in an acrobatic leaping into the air. It only survived in Japan into the latter part of the nineteenth century because the shogun had placed severe restrictions on who could bear guns, but after the relaxing of curbs on firearms, hunters soon extirpated this large and vulnerable creature. By the end of the century *Grus japonensis*, to give it its scientific name (meaning 'Japanese Crane') was thought wiped out from the Land of the Rising Sun, though there were still small populations in mainland Asia.

But then it was discovered again, half a century later, in an obscure area of Japan's northern island of Hokkaido. This time the government protected it, but in 1952 schoolchildren noticed that the cranes could not reach their usual food because unusually cold weather had frozen their feeding grounds. They gave the cranes corn, and the tradition of winter feeding continues to this day. Now they are a major tourist attraction – their charms boosted by the fact that the flock has grown from thirty when discovered to about 900 now.

The near-extinction of the Red-Crowned Crane in Japan is ironic, when one thinks that it had for centuries been a symbol of longevity in Japan, thought to live for a thousand years – and even in real life it can live for up to forty years, which is a long time for a bird. Japanese brides wore kimonos adorned with crane motifs, in the hope that their marriages would be long-lived too. But that didn't stop unsentimental hunters from carrying out their wholesale destruction.

The longevity myth lived on, however, and even found new life. A young girl fatally wounded by the Hiroshima atom bomb vowed before she died that she would make a thousand paper cranes. Although she passed away before she could finish the task, other children took up her wish, and even today Japanese children use origami to adorn schools with a thousand interlocked cranes.

In North America, Whooping Cranes almost became extinct too, falling to twenty-one birds by 1941. Captive breeding programmes have increased their numbers to about 500, after conservationists learnt lessons from earlier false starts during attempts to preserve other birds. Some of these false starts were distinctly comic. Many of the Whooping Cranes were first fostered by the much commoner Sandhill Crane – maximising the survival rate for baby birds by artificially increasing the number of parents. However, the fostered Whooping Cranes grew up with an identity crisis. They thought they were Sandhill Cranes too, and refused to mate with their own kind. In later programmes humans took on the task of rearing the young, disguising themselves by dressing in Whooping Crane costumes with arm puppets that resembled their neck and head. Whooping Cranes reared in captivity also had no sense of a need to migrate, unlike wild cranes, which had learned the technique from their elders. The 'solution' was to dress up a human in white like a Whooping Crane, seat him or her precariously in a rather fragile-looking ultra-light aircraft, and persuade a group of young birds to follow in the contraption's wake as it wended its way along the cranes' migration route.

The beauty of cranes has fascinated every culture that has come into contact with them. The Ancient Greeks thought the flight of the Common Crane, which travels in a wedge formation like geese, gave Hermes, messenger of the gods, the inspiration to create the letters of the alphabet. Watching cranes has prompted us to create a verb and a noun apiece: if you crane your neck you can see many cranes busily engaged in construction work in our cities. It is also the origin of a common British surname, given to people who were tall and thin or with long legs. Cranes' rather human appearance, when compared with other birds, added an extra spark of interest, as well as some tall tales. Some experts think the modern-day legend of the 'Mothman' of West Virginia, which inspired the 2002 Hollywood movie *The Mothman Prophecies*, was based on a misidentification of a Sandhill Crane.

Although it would be an understatement to say I'm sceptical about Mothman, I have to admit that cranes can look remarkably like humans as they walk slowly up and down fields showing off their beauty – rather like good-looking teenagers in warm Mediterranean countries who spend their evenings walking round the town square with a 'look how beautiful I am' expression. The seventeenth-century Japanese haiku poet Enomoto Kikaku caught the cranes' manner perfectly, reflecting:

On New Year's dawn
Sedately, the cranes
Pace up and down

In England, the Common Crane became extinct in the sixteenth century, but only after leaving its mark in many place names to a rare degree for birds – in the Cranfords of Devon, Northamptonshire and even Hounslow, for example. It is also probably the commonest bird in European heraldry, because its long neck and erect posture gave it a reputation for vigilance to which noble families, determined to protect their land, money and male line, very much aspired. British ornithologists predicted in the late twentieth century that it might return to Britain as marshlands were finally protected and even restored, and their long vigil was rewarded. It can again be found for real in the marshland of East Anglia, having returned there in 1981.

Jean Sibelius, the Finnish composer, never lost his fascination with nature in general and cranes in particular. Two days before he died at the crane-like age of ninety-one, he returned from his morning walk to tell his wife Aino that a flock was on its way. 'There they come, the birds of my youth,' he said, looking up at the sky. One of the birds broke away from the group to circle above his home, as if acknowledging him. Two days later he died.

GAME BIRDS

GREAT BUSTARD

New colonists in times of boom and bust

The standard image of England in its natural state before the depredations of heavy urbanisation is of mile upon mile of fields, neatly and picturesquely divided by hedgerows.

In fact England has gone through several different physical incarnations, and the era of small and tidy fields was merely one of them. It was the Enclosure Period, which peaked in the eighteenth and (early) nineteenth centuries, that parcelled up space into this patchwork pattern, as more and more agricultural land became private property rather than acreage for common use. For hundreds of years before that, England was largely a country of wide open expanses, though even this had not existed forever, since until the arrival of agriculture the country was mainly forest and marshland.

The Great Bustard, a magnificent, arrogant, strutting giant of a bird that requires open country so that it can spy danger from a great distance, therefore dates not from the age before us, but the age before that. It is small wonder, then, that it became extinct in Britain back in the 1830s, though it survives in the names of three pubs near its old haunts and in the coats of arms of two counties.

The bird, which looks like an exotic russet-coloured turkey, can be excused for its extreme wariness. One reason for its disappearance in Britain was the increasing lack of suitable habitat. Another was that it was frequently shot, since it had the bad luck to be an extremely large and tasty feast, often served to the nobility. I have not personally eaten a bustard, but let names be our guide to their culinary value. The Latin scientific name of one of the Great Bustard's relatives, the Bengal Florican, was *Otis deliciosa* until it was renamed *Otis bengalensis* in an attempt to create a more politically correct moniker.

It is just as memorable alive in a field as on a dinner table. Male bustards are known for their spectacular courting displays. The male Lesser Florican, a bustard found in India, jumps up to 2 metres into the air as many as 500 times a day, emitting a loud rattling sound. That may not impress you – and I'm not sure it impresses me either, since I used to have a rather nervy and irritable boss who did more or less the same – but it has left a sufficient mark on females to secure the species' survival for millions of years. Most other bustard species work similarly hard at their courting. A Houbara Bustard's athletic display in Morocco once confused a British bird guide so much that he

notoriously first misidentified it as two dogs fighting and then as an Arab on a bicycle – the white crest mistaken for a burnous. The Great Bustard goes for the opposite approach, puffing up and sticking out various parts of its body, but then remaining stationary for long periods, posing like a bodybuilder in a competition. It is not perhaps surprising that such striking birds are among the earliest ever depicted in art, in cave paintings in Spain that are at least 6,000 years old. The drawings are crude – in one of them this bulky bird looks more like a Diplodocus than a bird – but by a process of elimination, we can conclude that they must be bustards.

Who would not want to reintroduce such a magnificent bird to Britain? There have been three attempts so far, but their history shows that it is far harder to re-establish a bird than to stop it from being wiped out in the first place. The first two attempts, in Norfolk in 1900 and in Wiltshire in the 1970s, failed. It is still too early to tell whether the most recent attempt, in Wiltshire again in 2004, will succeed.

How are the auspices for British reintroductions in general? The answer is that re-establishment is a chancy business, since it is difficult for scientists to work out exactly what the right habitat and conditions are for a species suddenly transplanted into the British countryside after such a long period. All that scientists can do is to maximise the chances of success, by studying the ecology of the bird in countries where it is successful in an effort to replicate it in Britain. We can say, at least, that introductions have worked before: the capercaillie was successfully put back into Scotland after becoming extinct there in the eighteenth century because of hunting and forest clearance, and the White-Tailed Eagle was successfully returned to Scotland in the 1970s, though this came after two failed attempts in the previous two decades. But a plan to recolonise it in East Anglia was cancelled in 2010 after the government decided to rein in spending because of the credit crunch. So if the Great Bustard is the British bird of (two) ages past, the White-Tailed Eagle is the most modern of all birds – a victim of present economies just as much as taxpayers, public services and government workers.

RED GROUSE

The Grouse Economy

Wouldn't it be wonderful if the powers that be ran huge parts of an entire country for the sake of one bird?

Yes and no, history suggests.

The Red Grouse, a rather fat-looking reddish-brown bird with a showy scarlet crown on the male that looks like an eyebrow raised in disapproval (and you will not blame it after reading about all it has been through), was at one time thought to be the only bird species restricted to Britain. Later on it unluckily bucked the global mania for splitting up different races of the same birds into full species. The Red Grouse is now considered merely the endemic British version of the globally widespread Willow Grouse, so it is probably a good thing that the robin in the end beat the Red Grouse in a national competition to find Britain's favourite bird, after readers of *The Times* threw their weight behind the robin's cause. If the Red Grouse had won, it would be rather embarrassing to explain to a foreigner that it had all been a silly misunderstanding.

But the affection in which the bird has been held historically was only partly based on its mistaken uniqueness. The biggest reason is that the grouse has been a key part of British cultural life for centuries. In the twenty-first century its influence is waning, but in Victorian times every red-blooded male toff used to get a tingling feeling in the days before 12 August. The Glorious Twelfth was the time to head up to the grouse moors of Scotland, because that was when the shooting season started.

Large swathes of the Highlands of Scotland were based on the Grouse Economy, which employed thousands of people to tend the moors and monitor the grouse even outside the season. During the season – which began after the grouses had bred, in order to preserve their numbers – local people had to be hired as beaters and general factotums for the upper-class shooters. (For another example of how economies can be supported by birds, see the Edible-Nest Swiftlet essay.)

The irony for the grouse was the same as for all well-managed game birds. If people wanted to shoot them, numbers had to be kept up – so the sport of killing allowed grouse numbers to thrive. There were up to five million before each season began, at the historical peak of the grouse's fortunes. There are only about 500,000 now, because grouse shooting is much rarer. The Grouse Economy also kept tracts of land undeveloped, to preserve them as rural habitat for grouse.

But there was a toll. The biggest side effect for birds was the destruction of raptors – birds of prey – because it was believed they rivalled humans by killing the grouse in huge numbers. Hence the extinction of the osprey and White-Tailed Eagle from Britain, although both are now back with us. Britons' mania, until quite recent times, for regarding birds of prey as vermin to be extirpated explains why, when I was a child, I used to wonder why we saw so few of them in Britain and so many on our holidays to mainland Europe. Only recently have raptor numbers staged a recovery.

Moreover, grouse moor is not the richest habitat for birds. A country walk in Dartmoor (where they were introduced) may yield Red Grouse, wheatear and stonechat, but not much else. Woodland and farmland, which if managed well allow a greater number of species to live, were doubtless removed to make way for such moors.

Although grouse shooting was described as a sport, there was something very unsporting about the road it eventually went down. By mid-Victorian times beaters were employed to find the birds and flush them out into the open, where they could be shot. Purists attacked this, arguing that the true sport lay in knowing one's quarry well enough to find and then catch it – a process that could take hours.

There was something sinister about the mathematics of grouse shooting by late Victorian times – a casual indifference to life that presages the annihilation of World War One. The record was 1,070 grouse shot in a single day, by Lord Walsingham in 1888.

But the destruction of birds of prey on behalf of the grouse was based on false premises. In Edwardian times it was discovered – by a parliamentary committee, no less – that the chief cause of Red Grouse mortality in fact lay within the grouse itself. It was the parasitic threadworm *Trichostrongylus tenuis*, which kills grouse if the number within any single bird rises high enough. So a lot of birds of prey died for no purpose – beware of persecuting false enemies.

Does the extermination of the Red Grouse as a separate species leave us with nothing to call our own? This is another case of yes and no. Some scientists have now decided that one race of the Common Crossbill, a kind of finch with, literally, a bill that crosses at the end, merits the honour of being called a separate species, Scottish Crossbill. But other scientists disagree, saying the bird is merely a race of the Common Crossbill or even of the Parrot Crossbill, named for its particularly large beak. It's enough to make a Scot turn to whisky – Famous Grouse for anyone?

RED-LEGGED PARTRIDGE

The world lister's biggest supporter

The Red-Legged Partridge owes its origin in Britain to nothing less than royal dispensation, because Charles II introduced this fine-looking bird from Continental Europe for shooting in the seventeenth century. It has a white face, red eye patch and red bill that stand out against the earthen brown fields that it favours. After further introductions over the next hundred years or so it became fully established and now far outnumbers our native Grey Partridge. The latter, though rather rotund, can nevertheless often be found running rather fast along fields by the roadside.

But the Red-Legged Partridge has had a tough time inveigling itself into the affections of Britons, a conservative bunch who can take hundreds of years to get used to a foreign interloper on their land, and who suspect the decline of the Grey is linked to 'fowl play' by the Red-Legged – though there is no clear link between the rise of one and the fall of the other.

The Red-Legged Partridge is also, probably, the celebrity partridge that sits alone in a pear tree throughout the 'Twelve Days of Christmas' carol. Naturalists plump for this bird above Britain's native Grey Partridge for the common-sense reason that Grey Partridges don't like sitting in trees, whilst Red-Leggeds do – at least sometimes.

The killing of the bird was the origin of the success of Tom Gullick, the world's current top world lister – the man who has seen the highest number of bird species in the world (about 8,800 out of 10,000). Leaving the navy for a whirlwind career in which he helped to create Britain's embryonic package tour industry in the 1960s, Gullick then started a third new career in Spain – running Red-Legged Partridge shoots. This financed his world-listing trips abroad, which took up about two months each year.

Certain male birdwatchers are obsessed with their lists – with ticking new birds off, counting the number seen either in Britain or in the world or both, and then comparing their tallies with other people's. A variant of this is 'bird races', where teams get up a long time before the crack of dawn to vie for which can see or hear the most species in a calendar day. The most favoured time and place in Britain is Norfolk in May, when the migrant birds can easily be heard singing to attract mates, and a few migrants that shouldn't be here but have strayed can be added to lists. Norfolk offers the most birds of any British county in pretty much any month of the year, but careful calculations are sometimes made about whether it is worth starting in Yorkshire at midnight to hear the wailing of a few seabirds such as guillemots, before hurtling down the motorway to East Anglia, accompanied by the smell of burning rubber, in time for dawn. The current British bird race record is more than 160.

These events can become quite competitive and even fractious at times, as teams engage in practices of high or low cunning. Bill Oddie, who took part in a bird race in the 1980s, was miffed to see his team lose by only one bird, to a rival team that had been slightly late in reaching the arranged finishing line at midnight because they had driven off to listen for a Spotted Crake – a secretive reed-dwelling bird heard far more than seen.

People – and particularly long-suffering females of the species – speculate about this male obsession with bird lists. One theory aired by scientists and taken up by the birding equivalent of golfing widows is that all men are slightly autistic, and this is a blessedly mild form of autistic behaviour. Another is that ticking birds off, as opposed to simply watching their behaviour with scientific interest or enjoying them aesthetically, is a sublimated form of the male hunting instinct.

Gullick, who has become a minor birding celebrity because of his record, is quite clear when asked about what drives him: 'My motivation is very much a sublimated hunting instinct, but it's not all that sublimated. In fact it's very strong, since I also run partridge shoots.' Questioned about the joy of seeing the beauty of birds, he has said: 'I don't think there's much aesthetic pleasure about it. The pleasure is in the challenge and the achievement, the thrill when you see the bird.'

So what is the strategy required to be a champion world lister? You need a fair amount of money to get to all the obscure places in which birds of restricted range are found. Many world listers also suggest you should concentrate on one continent at a time, because you won't be able to remember what all the birds in every part of the world look like. You also need good local guides, because they will always have the edge on you in finding a particular bird. And then, of course, there are some skills that you're either born with or not. World listers need a good memory not only for birds, but also for people and places.

World listing is still pretty new. It only really began in the 1970s, practised by a handful of people. But at one point it seemed the next road that birdwatching might go down – the ornithological expression of the Global Village phenomenon much discussed by the media in the 1990s. However, the rise of the Green movement and the parallel yen for carbon-neutral lives have probably put paid to that. Jetting round the world for birds creates a heck of a lot of carbon emissions. World listing may prove to be a temporary phenomenon that existed for the short period in history when nature-loving people had become more internationally minded, but had not yet developed the international environmental awareness that shortly followed it.

Gullick's favourite bird? He cites the Red-Legged Partridge, the partridge that laid his golden eggs.

SONGSTERS

NIGHTINGALE

Sweet delights in a plain brown wrapper

Is there a bird in the world that is at the same time more irritating and more endearing than the nightingale?

First, let's dwell on the nightingale's good qualities. The male nightingale – for as with most other birds, it is the male who does the singing – is perhaps the finest songbird in the world. The sound is not happy (like a skylark's), operatically powerful (like a blackcap's), or persistent (like a blackbird's). What it instead excels in is an endless set of variations on a rather melancholy theme. No snatch of nightingale song is exactly the same as what came ten minutes before, or will come ten minutes after.

The nightingale's song is associated with romantic love, and it is easy to see why. It captures the bittersweet, pining qualities of this curious state of mind. The nightingale's habit of singing at night when most other birds are silent also makes it a bird for lovers, at a time for lovers.

John Clare, the English poet and early birdwatcher of the nineteenth century, noticed the romantic effect which the nightingale had on its compatriots. He once observed a couple 'lavishing praises on the beautiful song of the Nightingale'. This would be a moving tale indeed, except that Clare, who knew birdsong well, lamented that the bird was actually a thrush. He was probably not the first person to have complained that people have fancifully heard nightingales (and felt romance) for centuries where none were to be had, and he was certainly not the last.

Though a great poet, Clare seems to have been rather a cantankerous old sort. He also criticised the poetry of John Keats, author of the famous 1819 'Ode to a Nightingale', for describing nature 'as she appeared to his fancies' rather than in reality. But Keats did, to be fair to Clare, commit a grave ornithological error in his ode, by describing the bird singing as it flew away. Nightingales never sing while in flight – that would be far too helpful for birdwatchers trying to find them.

This leads on to their irritating side. Nightingales are extremely hard to see. Many an hour has been spent by birdwatchers staring at a bush waiting for them

to appear. This skulking behaviour is not perhaps unusual – many other birds indulge in it and some are even worse (just try seeing an Andalusian Hemipode, even in Andalusia). But what is exceptional about nightingales is that even during the spring courting season, when they are trying to attract a mate, they don't sing that often. Cynics have expressed incredulity at the celebrated live duets, during the early days of BBC Radio, between the cellist Beatrice Harrison and a nightingale. Although by all accounts the innovative first recording in 1924 left a deep and unifying impression on the nation – people kept their front doors open so that neighbours could hear the performance – it has been suggested that they are simply not sufficiently regular singers to be relied on. Never work with children or animals, as they say in showbiz – and never work with nightingales either. They are the prima donnas of the bird world who, showing a stubbornness that recalls Maria Callas' notorious unreliability, can never be completely depended on to turn up and sing. On top of their secretive tendency, they are also rare, with fewer than 7,000 pairs in Britain compared with up to a million in some European countries.

But of course this elusiveness is partly why they are so prized in Britain. To hear a nightingale is a difficult thing, and therefore a great prize to be treasured. It is, also, the most important thing – more important than seeing the rather drab dark brown body, distinguished only by a fetching reddish tail. Clare wryly notes this low return on the considerable investment in time required for seeing one. In his own ode to the bird, 'The Nightingale's Nest', he admits, like many a modern nature lover, that 'I've searched about/For hours in vain' to find it. After doing so, he concludes that:

> […] her renown
> Hath made me marvel that so famed a bird
> Should have no better dress than russet brown.

But he was sufficiently enamoured of the bird to write a little flock of poems about it.

However, despite Keats' ornithological shortcomings, his 'Ode to a Nightingale' is the best poem. He makes a perceptive point about the appreciation of nature: that it is a timeless and deeply democratic pleasure. The best things in life are free and as old as the hills, and that includes birdsong. In the words of Keats:

> The voice I hear this passing night was heard
> In ancient days by emperor and clown

If it is a poem as much about the universal equality of humanity as it is about romance, perhaps it is appropriate after all that it was written, rather unromantically, in a pub garden.

LINNET

Better than Wordsworth?

The male linnet is a handsome, rather dainty finch, with a red breast and small red crown when in breeding plumage, and an attractive meandering, twittering call like the singing of a vocalist in an experimental jazz band. It is still relatively common in Britain, despite a moderate decline in recent decades caused by the country's mania for needlessly tidying up the landscape by trimming hedges, removing scrub, and generally depriving it of the combination of open country and low vegetation often found on the edge of woodland, which it likes (for more on how excessive tidiness is bad for birds, read the Swift essay). But be warned: if you approach for a closer look at the bird's rather subtle streaked plumage, it will bolt with that bouncy flight that finches have, which makes it look as if they are just about to land at any moment – though this is a false hope, since they usually don't. The linnet is a nervous little creature.

This winning combination of physical beauty and songster's skill has made it a popular cage bird over the years. In the old music hall song 'My Old Man', about a family doing a midnight runner from its home to avoid paying the rent, the wife follows behind 'wiv me old cock linnet.' On a more rarefied plane, the bird has also garnered much poetic attention over the years, but the highlight of the linnet's life as artist's muse must surely be the time of William Wordsworth's fascination with the bird.

> Books! 'tis a dull and endless strife:
> Come, hear the woodland linnet,
> How sweet his music! on my life,
> There's more of wisdom in it.

So writes Wordsworth in 'The Tables Turned', a 1798 diatribe warning against the perils of civilisation and calling instead for a life in closer harmony with nature.

It is an alluring prospect: that we will achieve more wisdom on a woodland walk, surrounded by the inspiring sounds of nature, than from hundreds of hours stooped over our books, trying to educate ourselves to gain an advantage in the human rat race for money and status, or what Wordsworth summed up nattily in another poem as 'getting and spending'. Instead, in 'The Tables Turned', Wordsworth offers us the 'ready wealth' of nature – the joy, which no money can buy, of birdsong and of nature in general. To him the real wisdom of life lies in sticking to nature, rather than overindulging in a life of education and acquisition, like children of the twenty-first century pushed by their parents to do well at school so they can go to the right university and become wealthy bankers.

But cynical children schooled in Wordsworth have for the past two centuries commented on the poet's paradox: if nature is so great, shouldn't they be playing outside, surrounded by the song of the linnet, rather than stuck indoors reading Wordsworth's poems?

Many nature-loving poets have, to be fair, been acutely aware of the inability of human language to convey in its full glory the beauty of nature, and particularly of birdsong. The Romans served nightingales' tongues at banquets to emphasise their mastery even over one of the world's greatest songsters, but this merely emphasised that the nightingale had a gift for melody that few humans possessed. Shelley's 1820 'To a Skylark', written a millennium and a half later, vies with Keats' 'Ode To a Nightingale' as the best poem ever written celebrating the joy of birdsong, but Shelley's poem still takes a moment to rue the fact that the bird can convey the emotion of joy better than any poet – including, of course, a poet writing about the skylark. Robert Bridges, born in 1844, six years before Wordsworth's death in 1850, carried on the Lakeland poet's tradition of nature verse but devoted much of his short four-stanza poem on 'The Linnet' to worrying that his 'distorting' verse was not doing full justice to the beauty of the bird's song. Less tortured readers who think that *Angst* is a German centre-forward rather than a state of mind might feel like saying, 'Get over it and make your best fist of describing the bird – we can make up our own minds about whether you've messed it up or not.' But it is a more doubt-ridden poem written in a more doubt-ridden age – showing how our interpretation of birdsong reflects the temper of the times.

Society's darkening mood explains why birdsong had become a symbol not of joy but of sorrow by the time the US poet Walt Whitman completed his 'Memories of President Lincoln' almost a century after Wordsworth's poem. Whitman had seen at first hand the American Civil War, one of the first modern wars in history, characterised before its close by the daily horrors of trench warfare. The Hermit Thrush actually has a pleasant fluting song similar to other members of the thrush family, but in this poem it was a 'song of the bleeding throat'.

Perhaps the best known poem on birdsong published in the past century strikes a medium (on the happy side) between Wordsworth's panegyrics for birdsong and Whitman's pessimism, by suggesting that birdsong can console after all – if only for a moment. Writing about a train trip taken in June 1914 on the eve of the Great War, Edward Thomas remembered with affection an instant of peace at a brief unscheduled stop in the obscure Cotswold village of Adlestrop, when:

[...] for that minute a blackbird sang
Close by, and round him, mistier,
Farther and farther, all the birds
of Oxfordshire and Gloucestershire.

Fans of Wordsworth will be pleased to note that those birds may well have included linnets, for which the bushes around the now abandoned station seem ideal territory.

BLACK REDSTART

Feasting on humanity's misfortunes

We have prospered by wreaking catastrophe on some birds, but other birds have in equal measure thrived on our ruination at different points in history.

The male Black Redstart is a bird of cold black plumage that breeds in cold grey places, with a rather pretty but very short and clipped song, reminiscent of an emotionally repressed dunnock (one of our common garden birds). Its one concession to colour is a constantly quivering red tail – 'start' is an old English word for this appendage. One of the easiest places to see a Black Redstart in Britain is on the outer walls of Dungeness nuclear power station in Kent – if you actually want to go there. The power station itself sits in the middle of a bleak expanse of shingle which ranks as one of the largest in the world, and walking across it is like taking one step forward and two steps back with each stride. When you grow tired of watching the Black Redstart, you can go and watch the gulls and terns feeding precisely where the power station discharges its heated water, if you can make it across the perfidious shingle. The temperature attracts the fish, and the fish attract the birds.

The Black Redstart has a taste for breeding in places shunned by most of humanity. It first established itself in Britain in 1942, nesting in parts of cities devastated by the Blitz. They were pretty similar to the rocky habitat it chose in mainland Europe.

The avocet, an elegant black-and-white wading bird that takes its name from 'advocate' because of the black caps that lawyers used to wear, also had a good war. Wiped out from Britain because of habitat loss by about 1840, it first restarted regular breeding in Britain in marshland flooded by the military to prevent German invasion, which proved the perfect habitat.

Even the Exclusion Zone around Chernobyl nuclear power station in Ukraine became a haven for birds after humans were evacuated in 1986. In particular, the Eagle Owl – a magnificent bird of prey far bigger than any of our own owls – prospers in the area.

It says little for humanity if birds can only feed on our misfortunes, like a vulture feasting on carrion – and we know that vultures gorged themselves

well on the corpses at Isandlwana in South Africa, the site of one of Britain's greatest military disasters in 1879. But can certain birds only thrive in areas we've been forced to neglect and allow to run wild because of war or catastrophe?

We criticise local people in poor countries for hunting endangered species, but forget they are only doing so because they are hungry and do not have enough money to feed themselves without catching some of their own prey. If local conservationists asked them to go and catch another, more common bird that is just as meaty, they would be met with incomprehension at best. Conversely, although greater material wealth is sometimes based on the unsustainable plunder of the earth's resources, including concreting over good bird habitat, affluence sometimes allows people to cultivate a greater interest in conservation. Once our basic needs are met, we can think about other creatures. The United States and northern Europe are relatively wealthy, and lead the world in nature conservation – their nature reserves paid for by public donations. People from these wealthy countries also help conservation more directly, by putting out food for the birds in winter.

China, whose economic growth is veritably rampant, is in that tricky awkward stage that is probably the worst of all for birds. If China is faced with a choice between preserving birds and industrial development, the birds almost always lose. The construction of the huge Three Gorges Dam on the Yangtze has drained the wetland that much of the world's population of endangered Siberian Cranes used, as well as driving the final nail into the coffin of the now probably extinct Yangtze River Dolphin. China could point, in its defence, to its protection of another crane, the Black-Necked, whose wintering grounds at Cao Hai were flooded to meet the cranes' needs and turned into a nature reserve. But it doesn't bode well that this was a cast-off – the land was given back to the cranes because attempts to turn it into farmland were unsuccessful.

Even in China, though, there are hopeful signs. Hong Kong's birdlife used to be recorded in fine-grained detail by Western expatriate birdwatchers.

The region's expat population is now dwindling, but the number of Chinese birdwatchers there is growing, and taking their place – suggesting that parts of China are creating the affluence that encourages birdwatching as a hobby.

However, the billion-dollar question remains as to whether the whole world can make itself wealthy without ruining the climate by warming up the earth. This would be the worst thing of all for birds – even for the gulls and terns that like the hot water flowing out of Dungeness power station, close to where the Black Redstart flicks its tail like a bird with a nervous tick, anxiously contemplating the future that humanity holds in store for it.

SONG THRUSH

If it's worth singing once...

The Song Thrush is the opposite of China's ostensibly beautiful but quietly fraying Forbidden City – it is not much to look at until you get close to it. This thrush has a dull brown back, but near at hand you can see the rather attractive inky black blotches on its breast. It looks as if the paint that has been forbidden in the Forbidden City has been daubed on the Song Thrush's breast instead.

The bird also has the odd habit of (loudly) singing everything twice. Just to repeat that point to make it crystal clear, it sings everything twice. The Victorian poet Robert Browning, in his poem 'Home-Thoughts, From Abroad', found this endearing:

> That's the wise thrush: he sings each song twice over,
> Lest you think he never could recapture
> The first fine careless rapture!

Browning was merely the latest in a long line of British poets to be enthralled by the bird. In the eighteenth century Thomas Chatterton lamented, on the death of his lover, 'Sweet his tongue as the throstle's note', though one can only assume that his paramour, in common with the thrush and with some artistic characters of a rather egomaniacal tendency, had a habit of repeating what he was saying. If your own paramour is called Mavis, you might be interested to know that her name comes from an old word for the bird – celebrated in Robert Burns' poem 'Hark! the Mavis'. If you yourself are called Mavis, either in commemoration of the bird's song or for some more sentimental reason, be thankful, if you do not like your name, that at least that you are not called Thrush.

One could argue, in the Song Thrush's defence, that if something is worth singing, it is worth singing again. It is a beautiful sound: an ever-varying warble with the tone of a clarinet. But this leads to a conundrum: on fine spring mornings when the throstle is in full throttle, I think of the famous observation of Konrad Lorenz, the Austrian ornithologist, that birdsong is 'more beautiful than necessary' (to read about Lorenz's ingenious experiments to test birds' instincts, see the Herring Gull essay). Surely the male Song Thrush could advertise his presence to females in a more sparing manner? Singing risks attracting predators, and reduces feeding time, so why should he sing so noisily? Moreover, what is the evolutionary argument for having a song so elaborate and drawn out, since the time taken to sing it would surely attract yet more predatory attention?

Lorenz's riddle is answered by 'Zahavi's handicap', named after the Israeli scientist who came up with the concept in the 1970s. Zahavi's handicap, like many observations about the way birds behave, is a good example of how we can learn about ourselves from looking at birds. It is the bird equivalent to the human habit of conspicuous consumption. If you have an expensive Savile Row suit, you are proving that you're so wealthy you can afford to squander money – so you will have easily enough to spend on the upkeep of children produced by the mate you are trying to attract. For Song Thrushes, the bird with the finest, most developed and most varied song is likely to interest the females, because he is signalling that he is genetically so well equipped that he can afford the burden of spending all that time on learning how to sing well, and still find enough to eat for himself. Since he has good genes, the female will want to procure them for her children by mating with him. It is also what motivates some birds to develop long tail streamers that attract females but get in the way of normal life. The sea-going male Pomarine Skua has found the perfect solution to the dilemma – sometimes it simply removes its own particular Zahavi's handicap by breaking off some of its tail feathers with its beak after finding a mate (to read about some more dastardly skua habits, see the Great Skua essay).

Zahavi used the example of the peafowl. The male, better known as the peacock, has a ridiculously large tail that gets terribly in the way of everyday life – but that's the point. He also cited the intriguing case of the Arabian Babbler, which practises what he called 'competitive altruism'. These birds fall over themselves to volunteer for guard duty to protect the flock rather than go and find food, because it makes them look powerful.

Sceptics may argue that all Song Thrushes sound more or less the same, so they have reached a competitive dead end where none has the advantage over the other. Song Thrushes might counter-argue that all humans sound more or less the same. Doubtless they do to the Song Thrush, which will not be able to tell the mellifluous sounds of a television presenter from the barely coherent voice of a man (possibly the same man, if we are to believe gossip columns) who has had a few too many down the pub. In fact Song Thrushes' voices do vary in the same way that ours do – as do those of many other songbirds.

The basic rule of thumb is that older birds are more skilled singers than younger ones. Young birds take about a month to learn the song fully. But there is also much evidence that birds such as the Song Thrush, which have a wide range of phrases, gradually add to them with the passing years – if they survive that long, which the bulk of any songbird population doesn't.

Scientists are also discovering that birds have dialects, so the subspecies of the Song Thrush that lives on the Outer Hebrides is likely to sound different from its Sassenach counterparts. This is not entirely surprising, since it looks slightly different too – its back is darker than other Song Thrushes'. This implies that the bird has started the process of evolving into a different species, which reinforces its tendency to have a different song. Birds that live on small islands tend to have less sophisticated songs, probably because they hear fewer of their comrades of the same species than other birds do, so there is less learning material available.

It may sound like a flight of fancy that birds learn song from each other, but a memorable anecdote about Australia's Superb Lyrebird illustrates the truth of this. In the 1930s a farmer kept one as a pet, and used to play the flute to it (presumably because his flute-playing was only good enough for a captive audience). He was particularly keen on such hardy old standards as 'Keel Row', a Tyneside folk song transplanted to Australia by poor white settlers. The lyrebird started imitating the farmer, and other lyrebirds started imitating the pet. Seventy years later, Australian lyrebirds were singing 'Keel Row' and other well-known Geordie classics tens of miles away.

If birds on isolated islands are the dullards of birdsong, how do city birds, living in an environment at the extreme other end, fare? Very well indeed. The British media was shocked to discover recently that research showed birds in urban areas tend to sing louder (sometimes loud enough to break European Union health and safety regulations if they were human), in an effort to assure audibility above the greater background noise. Some of them also sing longer notes, at a higher pitch, to distinguish themselves from the kerfuffle around them. I'm not shocked by this at all. If birds are able to adapt to their environment in this way, my confidence is greater that they will be able to adapt in the future in other ways forced on them by humanity. In any case, birds that sing louder in a metropolis are no different in this respect from humans – compare the way a New Yorker speaks with someone living in the countryside. Any city-dwelling thrush that didn't sing loudly enough would be a schlep, as a resident of the Big Apple would say.

(EUROPEAN) ROBIN

A much-loved bully

The official British List records all the birds ever seen in the British Isles. At the time of writing there are 592 species on it, and every year one or two more are added – vagrants from Siberia, North America, and other far-flung places.

But the very first bird on the British List is the robin (more properly known as the European Robin) – beating the Common Crane by about forty years. We know from the annals of monks that St Serf, an obscure Scottish saint even by the standards of Scottish saints, had a pet robin at around AD 530 that was killed by his cruel pupils. It is the first historical record of any British bird. This top-billing status is fitting given that it is probably Britain's most popular bird, though it is ironic that the first ever mention of a robin in Britain involves the bird being bullied, since the robin is itself the foremost bully of the British bird world. Having been bested in AD 530, it has been making up for it ever since.

So why is this ruffian – declared Britain's favourite bird by popular vote in 1960 – so well loved? Naturally, if we knew more about the intimate lives of Britain's birds, we would probably have different favourites. Perhaps, for example, we would pick the gannet, which is ever faithful to its mate but is remembered more by us for its gargantuan appetite (a 'gannet' is an unaffectionate nickname for a greedy person), and hold it in greater affection than the robin.

But the robin is more gorgeously coloured than the gannet, a large seabird with a face reminiscent of a painted clown's. It is the prettiest of our common birds, because of its bright orange breast. In that case, why is its nickname Robin Redbreast rather than Robin Orangebreast? That's because we didn't have the colour orange when the moniker first appeared, since we didn't have oranges. The word 'orange' does not appear as a colour in Western languages until the sixteenth century.

Unusually for birds, the female is as brightly coloured as the male – probably because, unlike most birds, she keeps her own territory during the winter to ensure a steady food supply, and the orange gives her a prominence that helps to advertise this fact against intruders.

We also love it because it appears to love us – perching on the gardener's wheelbarrow, approaching us when we're on a picnic, and occasionally even coming into houses to say hello if the doors are open. However, it does this not out of any genuine affection, but because it thinks we're pigs, or some other large animal usefully digging up the earth to reveal earthworms which it can eat – which, if we're sufficiently assiduous gardeners, we are. In mainland European countries where there are wild boar, we can see robins following them around too as they scratch around in the dirt for grubs. It is classic parasitic behaviour.

The robin's association with Christmas also warms our hearts towards it. But there has been much speculation about why the robin, in particular, has such a starring role in the festivity. A simple aesthetic explanation is that its orange breast looks good on cards, standing out against snow that in England is usually as imaginary as the robin's friendliness. Another possibility is the bird's association with postmen – the bearers of the cards – who used to rejoice in the nickname of 'Robin' because of their bright red coats. It is also safe to say that European birds with prominent red or orange on them are often given an association with Christ, through legends that tell us the colour derives from his spilt blood. The robin is said to have sung into Christ's ear to comfort him on the cross, and the goldfinch and crossbill are linked to Jesus too through tales related to the crucifixion (for more on the goldfinch's Christian symbolism, see the Goldfinch essay).

So why is the robin such a bully? The origin is its obsessive defence of territory. As a result, it will threaten other birds sometimes bigger than itself, by scurrying around making clicking noises. That doesn't sound very frightening to us, but probably does if you're about the size of a robin. This explains why we never see them in groups larger than pairs (and even then, only in the breeding season) – they don't want other robins in the way competing for food.

The territorial urge also explains why the robin is such a keen singer. Other birds tend to warble mainly on spring mornings when they're attracting a mate. But we can hear the robin singing in the winter too, and even at night, greeting revellers leaving the pub at chucking-out time with a rather slow and melancholy, though beautiful, soliloquy that sounds a bit like a sad admonishment of our drunken behaviour. Wordsworth caught the song well, both as a poet and as an observant naturalist, writing in his poem 'The Trosachs' of:

> The pensive warbler of the ruddy breast
> That moral sweeten by a heaven-taught lay
> Lulling the year, with all its cares, to rest!

The robin's ubiquity and willingness to sing with a beautiful sadness at all times makes it a kind of poor man's nightingale, though its song lacks the variety of the latter – which explains why, as a subject for poetry, the nightingale has for centuries trumped the robin despite our affections for this feisty bird. But the robin's constant urge to sing – and to do so as loudly as possible – earned it a small footnote in history recently. When the new Conservative prime minister

David Cameron and Liberal Democrat deputy prime minister Nick Clegg held their historic meeting in the garden of 10 Downing Street to inject some much-needed bonhomie into the new coalition, BBC Radio listeners asked which loud bird was upstaging the two Great Men by drowning out their comments with its voice. It was, of course, a robin.

GOLDFINCH

The crucifixion bird

The English artist William Hogarth is best known for his excoriating satirical paintings exposing the contemporary morals – or lack of them – of eighteenth-century society.

But he was also capable of more touching works – and none more so than *The Graham Children*, a 1742 portrait in London's Tate Gallery that can only be truly appreciated if you know the symbolism of the different birds in classical painting.

To the modern aficionado of portraits, it looks a cosy, innocent scene – sweet, but nothing more than that. Four young children, including a baby full of the joys of life, are happily beaming away, while a cat eyes up a goldfinch in its cage. Cute.

But if you know that in the artistic symbolism of the era, small songbirds represented the souls of young children, the picture has a different meaning. At the time the picture was painted, the baby – represented by the goldfinch – was dead. This is not a portrait commissioned by parents smug at their fecundity, but an *in memoriam*.

The goldfinch is one of the most frequently depicted birds of classical art, and there is a good explanation for this, grounded in Christian iconography. The red on its face and its predilection for eating thistle seeds associate it with Christ's suffering on the cross, while bedecked in a crown of thorns. The legend is that it removed a thorn from Jesus' head as he carried his cross to Calvary, and as it did so Christ's blood stained its face.

The goldfinch's association with children and with Christ's passion make it a frequent prop in paintings of the Madonna with child – foretelling the crucifixion even as the infant Jesus is happily playing. In Raphael's *Madonna of the Goldfinch* (*c*.1505) it graduates to a starring role. A young John the Baptist hands his playmate Jesus the bird, as the Virgin Mary looks on lovingly. In Tiepolo's portrait of the same title (*c*.1760), as if armed with foreknowledge she averts her eyes from the sight of Jesus holding one of the birds, whose bloody red is all the more striking against the infant's pale flesh. In all of these paintings its likeness has been rendered well, in contrast to most attempts at portraying birds up to the early 1800s.

Another possible explanation for why so many artists painted it is that the goldfinch is simply among the prettiest of all birds that we see in our everyday lives, guaranteed to add a touch of colour to the drabbest scene. Carel Fabritius, the seventeenth-century Dutch artist, actually painted a portrait which featured the goldfinch as the sole subject – a rare honour for birds at the time.

If you cannot work out the appeal of goldfinches, look for them the next time you find yourself in an open-air car park – for some reason, goldfinches love them, and I can almost guarantee that you will find a small party flying around in their peculiarly parabolic way. These finches are very rarely on their own. Instead, you can hear their distinctive tinkling sounds in little choruses. Some say it sounds like water running in a stream, but to me it is more akin to a bunch of children giggling at a shared joke. Birds that seem as much at home in city streets as in the countryside, they have also taken a particular liking to TV aerials, on which they perch in the spring while making a euphonious trilling sound to advertise for a mate. People have long captured goldfinches to use as cage birds because of their beauty and song – in Thomas Hardy's 1886 *The Mayor of Casterbridge*, the eponymous character gives a goldfinch to his long-lost daughter, and the neglected bird's lonely death mirrors his own. Bird-fanciers used to cross-breed males of the species with female canaries in an effort to capture the best of both birds' voices. But I prefer goldfinches, like the souls of small children, to be free.

Our goldfinch, more properly known as the European Goldfinch, has a transatlantic cousin, the American Goldfinch, which proves the adage that less is more. The male American Goldfinch is almost entirely bright yellow in its breeding plumage. The European Goldfinch is mainly brown and black, but with an odd flash of bright colour (some red on the head, some yellow on the wings) that is all the more appreciated because sometimes you have to wait a couple of seconds to see it. The best things in life are worth waiting for, and that applies to goldfinches too – particularly because you won't have to wait too long.

SKYLARK

The happiest bird in the world

The male skylark is loved most for its warbling song, so I realise I risk inviting the mockery of all nature-loving Britons to say that it is far from a virtuoso singer.

The skylark makes a fine contrast with the nightingale. The latter utters a series of infinite, complex variations on a theme, with each phrase slightly different. The skylark simply flies to the top of the sky to sing a repetitive tune, no more than a few seconds long, over and over again.

It is not much to look at either. The bird is brown and streaked, to camouflage it perfectly with the fields of stubble that it likes to feed in, and its sole concession to physical ornament is a modest brown crest. When foraging on the ground, the skylark seems a very lowly bird in all senses. A creature that appears veritably to hate trees, this bird of unprepossessing, mud-coloured plumage spends much of its time clambering about in fields of a similar hue.

But when the skylark takes to the skies, what a bird it seems then. It can ascend at an amazing rate, frequently flying high enough to pass into invisibility. This rapid rise seems all the more miraculous, and therefore special, when you have just spent the last ten minutes watching it crawling about on the ground. The sense of exhilaration that a small bird can fly so high is accentuated by its penchant for airborne acrobatics in large open spaces, rather than over woodland or scrubland. Imbued with these habits, the bird seems to embody the very spirit of the terrain you are standing in, making the skylark the unmatchable symbol of freedom. We like birds partly because their ability to fly gives them a liberation that we lack – there are times in everyone's life when they would like to fly away, up into the air, from an unpleasant situation. The skylark encapsulates this desire for liberty.

But the key ingredient of the skylark's appeal is its song. You may be surprised to read this since I have spent two paragraphs finding fault with its capabilities as a songster. But the song's very limitations make it all the more appealing. The skylark is not a skilled performer, but the sound has the twin virtues of persistence and an apparent happiness. A child's singing is often bad, but sometimes more moving than a trained opera singer's because it springs out of a spontaneous sense of joy – sung while youngsters are larking about (or, to take an older form little used nowadays, while they are skylarking). Like some children, a skylark never seems to stop singing, to the point where you wonder when it is managing to catch its breath. Its voice transmits an intense jollity that no human performance could ever hope to emulate. It is this happiness on which the poet Shelley waxes lyrical in his 1820 ode 'To a Skylark':

What objects are the fountains
Of thy happy strain?
What fields, or waves, or mountains?
What shapes of sky or plain?
What love of thine own kind? What ignorance of pain?

For me the greatest joy lies not in seeing a skylark, but in hearing it singing from such a point that I can no longer see it. This more than makes up for the sense of embarrassed incompetence one always feels when not able to point out to non-birdwatchers a bird whose voice is so loud, and whose habitat is so unencumbered by trees and bushes, that it seems that only a fool could fail to find it. It also makes up for the slight sense of vertigo, mixed with aching, felt when straining to see a bird which seems to have flown straight up to heaven.

Birds' ability to fly has prompted humans to associate them with the heavens and hence with gods throughout history, and the skylark's fondness for climbing so high to sing has sharpened this association. Edmund Spenser wrote in the sixteenth century of how 'the merry Larke hir mattins sings aloft', in reference to the morning prayers sung by priests. A century later the poet John Dryden was likening the ascent of human souls to heaven on the day of the Last Judgement to 'mounting larks', in an ode to a deceased friend. Even Shelley, one of Western civilisation's first public atheists, had to resort to semi-religious language to describe the bird, addressing it as a 'blithe spirit'.

So why does the male skylark carol so high, and for so long? Evolutionary biologists say this is to show the female that it can. This may sound inherently wasteful – why don't perfectly evolved species instead spend their time on more practical things, like looking for food? But the skylark that can reach the highest and sing the longest is proving it is a fitter bird, and therefore has better genes.

This doesn't exclude the possibility that the skylark *is* actually happy when it warbles. It is not fanciful to suggest that evolution has favoured those skylarks with an urge to reach for the skies and sing their hearts out – and even the mere fulfilment of an urge produces some sense of satisfaction in a creature with a central nervous system. Evolution does not necessarily have to crowd out all romantic notions. In the same way, Shelley's atheism, a rarity at that point in history that has become common in modern Western society, did not destroy his pleasure at one of the little mysteries of nature: the skylark's ability to confer joy on all who hear it, whatever their religious views.

CETTI'S WARBLER

Was Beethoven a birdwatcher?

The greatest puzzle about the Cetti's Warbler is a musical one – did it inspire the opening notes of the last movement of Beethoven's Second Symphony?

The rhythm and tone of the opening part of the bird's staccato call – 'DUM da-da-da-DA-da' – are remarkably similar to the first flourish of Beethoven's fourth movement, giving both bird and orchestral work a slightly impudent timbre. The symphony, like the bird, seems to be saying to its audience: 'WHAT? So-why-are-YOU-here?'

The resemblance might be sheer coincidence, but the question has to be asked: was Beethoven a birdwatcher? Did he hear the Cetti's Warbler, and borrow it for his symphony?

Beethoven was certainly a nature lover who enjoyed rustic rambles, and there are occasions on which he unambiguously used birds in his music. His Sixth Symphony – the Pastoral – contains recognisable bird calls, including that of the nightingale (flute), the 'wet-my-lips' sound of the quail (oboe) and the call of the cuckoo (clarinet), probably the most common bird to feature in classical music. (For more on this mischievous disyllabic bird, see the Cuckoo essay.)

But when it comes to the Cetti's Warbler, my guess is that its transfer from swamp to symphony most likely lies somewhere in the twilight zone between chance and conscious imitation. One can imagine the striking sound lodged in Beethoven's brain on a country walk without his knowledge, and then committed to a musical score.

Those who see the origin of human music-making in the songs of birds would be wise to cite this composer, but he is not the only one to remind me of birdsong. The wren's long trills bear an uncanny resemblance to Maria Callas' showy *bel canto* singing in Donizetti operas of the early nineteenth century. In particular, there is a repeated phrase in *Tornami a dir che m'ami*, an aria from Don Pasquale, that sounds precisely likely the penultimate phrase of the wren's four-part song.

At a more fundamental level, we should bear in mind that the world's first music was made not by humans but by birds (and, arguably, also by whales, whose noises sound uncannily like birdsong when speeded up many, many times), long before humans invented musical instruments. It is far more likely

that humanity borrowed from the music it heard from birds when inventing its own, thousands of years ago, rather than simply ignoring nature's teaching and starting again from scratch.

The second great puzzle to the Cetti's Warbler (named after the eighteenth-century zoologist Francesco Cetti – pronounced 'Chetty') is the challenge of discovering what it actually looks like. A reddish tint to its brownish plumage pushes it over the boundary from dullness into subtle beauty, but you will be lucky to appreciate this: the warbler hides itself in dense vegetation and, in diametric opposition to an obedient Victorian child, is more often heard than seen.

Another riddle is why it can be seen in Britain at all. It is one of the country's newest breeding birds. Having gradually expanded northwards on the Continent, the Cetti's decided next to colonise the marshes of southern England in the 1970s. Most scientists had bet, instead, on the colonisation of the Zitting Cisticola, another marsh-dwelling bird that had moved northward. The presence of dramatic Beethovenian sounds from the reeds, and the absence of the cisticola's eponymous one-note 'zit' call, show the unpredictability of nature even for the most experienced ornithologists.

Having arrived here, the ill-mannered Cetti's Warbler is refusing to co-operate by showing itself to a public eager to see this relative novelty. It is immensely fond of teasing birdwatchers, by quickly rattling off its machine-gun-like song, scampering off somewhere else while birders try to locate the sound, and then repeating the same discombobulating trick from a discreet distance. Scientists call this the Beau Geste principle, after the fictional Foreign Legion hero who defended his fort by leaving dead comrades propped up on the battlements to give the impression of a greater number.

Such techniques are not merely the preserve of birds and the protagonists of seemingly improbable boys' fiction. In the American Civil War the Confederate general Robert E. Lee decided to keep manically moving his men around within sight of the enemy, to forestall attack by giving the impression that his outnumbered force was greater than it was. It worked: George McClellan, commander of the opposing Union forces, wired back to Washington DC saying he needed 100,000 more men before he could take the offensive. His frustrated boss, President Abraham Lincoln, soon found a replacement.

The victims of the Cetti's Warbler suspect that it, like General Lee, draws a mischievous pleasure from this highly effective habit. If you manage to see it, you will notice that it often holds its long tail cocked like a wren's, giving it the same soupçon of triumphant cheekiness – like a schoolboy breaking the school rules by keeping his cap at a jaunty angle. This cockiness seems appropriate: given the bird's successful elusiveness, it, not its audience, usually enjoys the last laugh.

BLACKCAP

Rapid-reaction evolution

I knew the blackcap was prospering in Britain when I saw one in the Kent garden of a friend – possibly the smallest garden in England – one winter's day.

I'm pleased that this hyperactive little bird is faring so well, and for several reasons.

Seeing a pair of blackcaps is almost like getting two species in one. The male has a striking black cap – hence the name – on a body that has the cold grey hue of gravestones. The female is, unusually for female birds, which tend to be rather drab, a distinctly jollier-looking version. Its cap is the sort of vivid rufous-brown that often graces the painted walls of modern living rooms – or, for the romantically inclined who treasure those imaginative names for paint colours listed in the free catalogues you get in DIY stores, I might describe it as a pleasant Hot Fudge or perhaps Autumn Shower. But birds have generally been named by men, so unsurprisingly the male's headwear rather than the female's inspired the name (for more on odd bird names, see the Arctic Tern essay).

Blackcaps also have a rich, fluting and exceptionally long-winded song. Those of a sunny (but patient) disposition often prefer it to the nightingale's rather melancholy strains, and the blackcap has even been called the 'northern nightingale' because it breeds further north (as well as on the same southern commons where the nightingale can be heard).

But the biggest reason why I rejoice in the blackcap's success is that it shows the amazing ability for birds to adapt to change, on the brink of an era when the ability to do so could decide whether many species go extinct or not.

A few decades ago we only saw blackcaps in the summer in Britain, when they came here to breed. They would then migrate to Spain and other points south for the winter. But then we started seeing blackcaps in the winter too, and rejoiced in the fact that some birds had decided to stick with us all year round. However, we were wrong: these were not the same individuals as the summer ones. Very little is straightforward in the bird world, and blackcaps are no exception. The winter birds bred in Germany, and the birds that bred in Britain continued to go south. The German blackcaps were, it seemed, responding to Britain's national obsession with putting out seeds for the birds to eat in the winter by coming here to live on them.

In 2009 German academics put out a revolutionary new paper, which found that in fewer than thirty generations, one group of German blackcaps had not only decided to travel to Britain for the winter to feed on bird tables – they had even evolved to do so. They now had shorter, rounder wings than the blackcaps that went south (whether from Britain or from Germany), because they had less far to migrate – and birds that don't migrate very far don't need the longer, straighter wings that make it easier to fly large distances. They also had longer, thinner bills – better for eating seeds on British bird tables, but no longer suitable for eating the fruits that blackcaps in Spain spend the winter eating.

Thirty generations may sound a long time – and for humans that might mean 700 years. But blackcaps produce a new generation every year, which is ready to breed a year later. Thirty years is not only a very short time in evolutionary terms, it is even a short time in human history.

What are the implications of this? Although the blackcaps' evolution was not in response to climate change, the rapid pace of its adaptation suggests that some birds may be speedy enough to respond to climate change after all. The phenomenon even suggests that humans may be able to give evolution a helping hand through the feeding opportunities they offer – which is good news, since this could help us to redress the harm we have caused through global warming.

The blackcap story is all the more amazing when we consider that the blackcaps that come to Britain in the colder months breed in the same German forests as other German blackcaps that winter in Spain. Clearly they do not intermingle, because if they did, a race of rounder-winged blackcaps would not have emerged. They probably don't mix together because the birds that go to Britain form pair bonds (ornithologist-speak for starting to step out with each other) in their winter quarters. They also probably return to Germany earlier, so have started breeding before the Germans which overwintered in Spain have returned. This raises a broader question, about which we know little: have other birds that have started spending the winter in Britain for similar reasons to the blackcap also started evolving separately – such as the chiffchaff, another warbler that, in parts of the south, can now be seen all year round?

But enough of all that – now it's time to launch a campaign to rename the bird after the female's crown colour, in the cause of gender equality. Where's that paint catalogue from B&Q? How about 'Autumn Showercap?'

WOODLARK

Did Burns bungle?

The woodlark – the lesser-known cousin of the skylark – is at the centre of a historical mystery story.

Robert Burns, Scotland's Shakespeare, wrote an 'Address to the Woodlark' that praised its beautiful song, and why not, you may ask? Poets are prone to write about birds and particularly their voices, and some naturalists credit the woodlark with the prettiest song in Britain – a mixture of trilling and yodelling, with many variations just like the nightingale's. Many ornithologists also think it has the added bonus of the most mellifluous of all Latin scientific names for a bird – the partly onomatopoeic *Lullula arborea*. In common too with the nightingale (for more on this bird lauded by poets over the centuries, see the Nightingale essay), its fine song more than compensates for its rather drab streaked plumage, ending in an inelegant, stumpy tail.

But there's just one problem, Mr Burns: there aren't any woodlarks in Scotland.

One possibility is that Burns, who never went further south than the very north of England, simply made a mistake. It's hardly likely he confused it with the happy skylark, since his poem dwells on the heartbreaking quality of the song:

> Thou tells o' never-ending care;
> O' speechless grief, and dark despair:
> For pity's sake, sweet bird, nae mair!
> Or my poor heart is broken.

But it is possible he confused it with the Tree Pipit, which is brown and streaked like the woodlark, lives in Scotland, and has a song which, though not positively drenched in melancholy, might seem so to a poet bent on a spot of wallowing.

But are we guilty of the modern sin of believing that the present age always gets its ornithological facts right, and that previous ages are always wrong if their observations clash with our own?

British birds' distribution has changed massively over the centuries. Remains of specimens show that we used to have pelicans in the Somerset levels in prehistoric times. The House Sparrow, one of our commonest birds, probably became so only with the growth of farming (for more on the House Sparrow's rise and fall, see the House Sparrow essay) – and so too, probably, did the skylark, the woodlark's closest cousin in England. There is also strong evidence that the Golden Oriole, a rare breeder now confined to East Anglia, was found as far west as Wales in medieval times. Its 'yellow colour and sweet whistle' were described in a twelfth-century manuscript by Giraldus Cambrensis, a Welshman who made many other astute observations about birds that are known to be accurate.

But past sightings are often mocked if they do not tally with our present experience. It has even been suggested that all birds found before 1959, when the official British Birds Rarities Committee was set up to accept or reject unusual sightings, should be wiped from the British list – as if nothing, and no bird, existed before then.

This ornithological book-burning would see expunged from the official British list several birds that have not been seen here for the past fifty-two years simply because human depredations have severely reduced their numbers, such as the North American Eskimo Curlew. Such an airbrushing of history would remove some ancient records whose presence on the list serves as a reminder of the need to conserve nature. More fundamentally, it would also be an insult to historical scholarship (for a salutary lesson on how it's good to keep an open mind about avian tales from the past, however improbable, see the Northern Bald Ibis essay).

But, returning to the Scottish poet, was he right or wrong? There are three arguments on Robbie's side:

- The first is that his writings show him in general to be an accurate observer of nature. So if he has good form when it comes to other birds, perhaps he is right about the woodlark.

- The second is that naturalists have found another, very believable, old reference to a woodlark in Scotland, from a priest who refers to its nocturnal autumn song. Woodlarks are well known – in contrast to Tree Pipits – for singing at these unusual times of the day and year.

- The third argument on Burns' side is that the woodlark declined rapidly as a species in England until the 1980s, so we don't really know where it was during its peak. It is currently found in eastern and southern England, but who knows where it will end up if numbers continue to rebound? If it eventually expands to Scotland, the future will prove that the past observation could have been correct.

Anyway, we could argue that it doesn't really matter whether Burns got it wrong or not. His poem on the woodlark is a splendid work of art. It sums up, for me, the bird's song much more effectively than the Sassenach poet Gerard Manley Hopkins' more direct attempt to do so by writing down the sounds he heard. Hopkins renders the woodlark's song as *Teevo cheevo cheevio chee* – a brave stab at pinning down birdsong into words, but one that fails to capture the woodlark's aural beauty. One shouldn't be too harsh on Hopkins though – even modern field guides do no better, when trying to describe the ineffable for the printed page (to understand just why writing down birdsong is so hard, see the Yellowhammer essay). Sometimes it is better just to go out and listen to the bird.

BLACKBIRD

Keeping live music alive

Who needs an introduction to the blackbird, fond of sitting on television aerials in many a street making a huge fuss about nothing? When disturbed by unwitting passers-by minding their own business, it abruptly rushes off with a long tut-tutting sound that has the same affronted quality as a hen's cluck.

There are some common birds that non-birdwatchers rarely notice (read about an exemplar of elusiveness in the Wren essay), but the jolly blackbird is not one of them. Its readiness to live wherever there are trees, coupled with its love of bustling noisily around its territory, make it one of the most familiar of all British birds.

But is it as simple as that? Why do we like the blackbird so much? I see the blackbird as more of a counter-culture hero than a respectable suburban neighbour.

The male's positively jet-black colour gives it a certain sinister beauty, and explains why Merle – the French for the bird – has been used as a name for both boys and girls, including Merle Oberon, the dark-skinned Anglo-Indian beauty who became a film star in the 1930s. Although it is a creature primarily of the day, its hue reminds artists of the dark and slightly edgy things in life

– in the same way as birds of dawn and dusk, such as the nightjar, are often invoked to set a mournful introduction for a poem. In the history of human culture the blackbird has been called into action more to illustrate despair, darkness and death than life's lighter side.

On this note, there is much speculation that 'Bye Bye Blackbird', the 1920s jazz standard and one of the great mystery songs whose lyrical meaning remains unclear to fans, is about a prostitute leaving her trade to go home to her mother. The opaque lyrics certainly bear the aroma of sadness and depression.

Even the universally known nursery rhyme about four-and-twenty blackbirds baked in a pie is one of those ancient children's songs of a rather brutal strain that might upset the sensitively minded child of the twenty-first century. Children of past eras would probably not have been excessively scared by it. They would have been much more used to the everyday death of animals butchered for the dinner table (what a modern parent would call 'local and sustainably sourced ingredients'), and vermin exterminated for the health of the house.

But when modern children hear that cooked blackbirds suddenly start singing, they invariably ask why, their parents invariably don't know – and then the children start crying, leaving parents baulking at whether they should be teaching the song in an age when they are more desperately than ever trying to protect their adored scions against the graphic violence on show at the click of a computer mouse or the squeeze of a television remote control button. That's even before the mewing children get to the end of the song, where a blackbird pecks off the maid's nose while she's faithfully fulfilling her laundry duties – though, like a Hollywood producer on such occasions, the rhyme's composer could doubtless claim that such seemingly gratuitous violence is an integral part of the plot.

The answer to the child's question is that the song refers to the Tudor vogue for putting the birds in culinary dishes as a kind of party trick – though, unlike in the song, in reality they were inserted after the pie was cooked.

This explains why the blackbirds' powerful voices, which tend to dominate the soundscape when birdwatchers are trying to listen for other birds, and would certainly have succeeded in upstaging the local wits at any Tudor party, remained in perfect working order.

The blackbird is also one of the few birds that has made the transition from classical poetry, written in less urbanised ages when poets heard birdsong more frequently, to pop music, the art form that has inherited classical poetry's role as the primary means by which clever young people put their bittersweet thoughts of love and loss into rhyming couplets. We should be grateful: birdsong's survival as an inspiration for pop poets is largely down to this particular, familiar bird, as at home on Abbey Road in London as in the countryside.

A mere half-century or so after Julian Grenfell associated the blackbird with death in 'Into Battle', a poem written in the classical tradition during World War One, the birds popped up again prominently in at least two classic pop songs – both associated, naturally, with the dark. The Beatles' 'Blackbird' actually features a recording of a 'Blackbird singing in the dead of night'. And there is the blackbird in 'An Cat Dubh', an early U2 song from 1980 – a rather darkly told tale about a teenager's shame in losing his virginity to a girl in a darkened room.

Some birdwatchers rail against the Beatles' decision to sing a song about the blackbird's nocturnal noise – arguing that they probably got confused with the robin, which does sing at night and can sound like a blackbird to the uninitiated. But although the blackbird does not sing 'in the dead of night', as the Beatles would have us believe, it is one of those few British songbirds that is quite noisy after dark – making a rather smug chuckling sound as it prepares to roost, as if pleased to have found somewhere suitable for a kip. The bird only reluctantly yields aural territory to nocturnal callers like the owl at the time when twilight gloom has turned into pitch-black night. So I think the band is probably merely guilty of a touch of poetic licence, rather than outright error.

It's not just, perhaps, the blackbird's colour that has made it a bohemian icon among modern singers – a John Lennon with wings. Its rather lazy and unstructured song has a distinctly jazzy rhythm to it, like a linnet's. Moreover, in an age where increasing numbers of us live in cities where we hear birdsong less than our ancestors (although more than we think, if we listen carefully), we can usually still hear the widespread blackbird, which serves to keep alive our appreciation of this particular gift of nature. We are in an era when the glee clubs of the middle classes have largely disappeared, to be replaced by much more convenient CDs and downloads. So the blackbird's jolly and slightly cheeky sound – always almost but not quite what the listener is expecting to hear, with each phrase very slightly different from the previous one, just to confound our expectations – is often the only genuinely live music we will hear all day.

WREN

The common rare bird

The wren is with us almost everywhere in Britain, but so rarely noticed by the general public that it is often thought quite rare.

People have found wrens hard to see for centuries, and the difficulty in observing them has led to all manner of confusion, which underlines how little we knew about birds until a few hundred years ago. Many English people in medieval times thought the wren was the female of the robin – and the feminine association lingered for centuries, both in the nickname 'Wrens' given to members of the Women's Royal Naval Service in two World Wars, and in 'Jenny Wren', which survives even today as a non-birdwatchers' name for the bird. The association between the two is found in the long version of the ancient nursery rhyme 'Who Killed Cock Robin?', when the unfortunate redbreast is killed by the sparrow with his bow and arrow on the eve of marrying the wren. It survived into Shakespearian times, when the playwright and poet John Webster composed a rather grisly dirge to:

Call for the robin-redbreast and the wren,
Since o'er shady groves they hover,

And with leaves and flowers do cover
The friendless bodies of unburied men.

It seemed an appropriate request for the wren in particular, which devotes much of its time to skulking in bushes where the sun doesn't shine.

But despite its elusiveness, the wren has at times been the commonest bird in Britain, though from decade to decade it vies for this position with the blackbird and chaffinch.

The wren is also unusually ubiquitous as well as abundant. As long as it has a little patch of rather scruffy vegetation, it seems content. I have heard a wren singing in a spindly bush in Parliament Square – symbolically the epicentre of urban Britain. When carrying out official surveys of breeding birds in the spring, I have found myself in a blissfully uninterrupted state of wrendom for hours at a time – able to hear a new wren marking out its territory while the previous wren's voice is still close.

Wrens have even made it to the distant Scottish island range of St Kilda, where isolation has produced a separate subspecies that is slightly larger than the normal kind, in the same way that St Kilda men were reputed to have their own genetic peculiarity – long toes that came in handy when climbing cliffs to catch seabirds.

It is surprising at first, perhaps, to think that a bird as small as the wren – our third smallest after the goldcrest and firecrest, though often incorrectly said to be the tiniest of all – has been seen as the king of the birds across various cultures. A European legend mentioned by the Ancient Greek philosopher Aristotle recounts a contest to decide who would hold this title. It was decided that the most fitting way was to pick the bird that could fly the highest. The eagle appeared to be a dead cert, but the tiny wren was hiding in its plumage, and jumped out to fly a little higher and clinch the title just when the eagle had reached its zenith. So the wren won by opposing power with guile.

This bird legend has common parallels in stories of humans, from ancient legends to modern films. It is enticing to think that we can all overturn the pecking order. The wren's cunning in the ancient tale is not so very different from the guile of jesters in Shakespeare plays who, while appearing laughably low in society, end up giving sage advice (usually unheeded) to the king. The wren's fabled ability to get ahead is also not dissimilar to that in Joseph Losey's 1960s film *The Servant*, where the devious hired help wrests control from his high-born master.

The cunning wren is one of those birds which is much easier to find once you have learned a couple of tricks. The first is to get to grips with its song – a very melodious and rather long warble, but with a hint of urban aggression to it. You will then be aware that you are hearing a wren several times a day, and can start looking for it. Keep an eye out – at about eye level or lower, rather than gazing upwards into the trees – for that small brown bird which you might mistake for a sparrow, were it not for a rather more whirring flight.

The wren is, when we get the occasional chance to inspect it closely, a highly attractive little bird. Its habit of cocking its tail – which has a fine-looking rufous hue – gives it a certain mien of raffishness. It has an aesthetically pleasing chequerboard pattern of square dots on its wings and tail that make it look a little like a bird illustrated in some centuries-old tapestry created painstakingly with a thousand weaves and stitches. It has the same aura of beauty as the tiny but intricate Anglo-Saxon broaches found in museums. If you live in Britain, I bet that on a fine spring morning there is a wren singing within a few roads of your home – and it's well worth seeking it.

OTHER PERCHING BIRDS

BARE-FACED BULBUL

You ain't seen everything yet

As we begin the third millennium AD the public often assumes that we have discovered just about everything on this earth, but naturalists know that's not true at all – even though we have been just about everywhere, the age of discovery is not yet over.

About 10,000 species of insect are found every year, although some are perhaps being refound as there is no comprehensive catalogue of the one to two million that have already been discovered. Estimates of the number left to locate range from two to twenty-five million.

When it comes to birds, matters are by this stage of history less chaotic. Aristotle, the first great scholar to write down his thoughts on birds, could count 140 in the fourth century BC. Western medieval writers could only find a similar number, reflecting the fact that scientific knowledge in general did not make much progress for almost a millennium and a half, with ornithology no exception to this rule. Even in 1701 John Ray, who overcame his humble origin as a blacksmith's son to become England's first great naturalist, listed only 500 bird species across the world despite the recent discovery of exotic avifauna as European countries started exploring and building empires. He boldly speculated that there might be 33 per cent more, but was nowhere near bold enough, since we have now found about 10,000. We can tell, by the rate of new discoveries, that we have located the bulk of the world's birds, though that number could be pushed up or down by a couple of thousand each way, depending on how we choose to classify species.

Despite this, a few new birds are still being found, and the annual number has settled during the 2000s into a steady flow of about five. You might ask how new discoveries can be possible. Birds are much more conspicuous than insects, or even than mammals; they're not microscopically sized, and even if they aren't brightly coloured, they are always moving about, and the human eye is designed to detect movement. Surely we ought to have seen them all by now?

One answer to this question is that many of the new birds are from South America, where inhospitable habitat has made it hard for humans to venture. South America's rainforests are impenetrable because they are dense, and such density supports an immense variety of insect and plant life, creating huge numbers of different niches for birds to exploit. Each niche tends to create a new species, often found only in one small patch of territory and nowhere else in the world. There are also, across the globe, a fair few shadowy birds of the night – owls and nightjars – which are difficult to locate. The Serendib Scops-Owl of Sri Lanka was first heard by its eventual discoverer, Deepal Warakagoda, in 1995. But although he knew the call was not that of any bird known to science, Warakagoda didn't manage to see it until 2001.

Once in a while a bird far unlike any other is found – repeating the excitement that European explorers must have felt when seeing the first penguins or emus. If I could pick any bird discovered over the past decade that wins marks just for being plain peculiar, it would be the Bare-Faced Bulbul of Laos, which had the barefaced cheek not to reveal itself to the world until 2009.

'Weird' is a more fitting epithet than 'beautiful' for this new discovery. The thrush-sized bulbul's face is completely free of feathers, apart from the very back of the crown, which has a profusion of grey ones. In the middle of this bare face is a large eye surrounded by a wide swathe of blue pigmentation, so it looks like a balding old transvestite wearing blue eyeshadow. Leaving aside the baldness and cross-dressing for a moment, a female friend reliably informs me that blue eyeshadow went out of fashion in the 1980s.

How had such a distinctive and not remotely shy bird managed to escape detection for so long? One reason was political – Laos had seen decades of instability, and had only become a safe country to visit relatively recently. Another was practical – the bird is found on steep limestone terrain that is very hard to traverse. But there was also an element of bad luck. One of the ornithologists who eventually described the bird to science had seen it fourteen years before. But he had been the only one to spot it, and after much mockery by his colleagues had shyly decided not to mention the bird in the official trip report – hiding his bulbul under a bushel. The gap between *thinking* you've found a new bird and *knowing* you've seen it can indeed be long, though for the Congo Peafowl it was even more extended than for the bulbul. Its existence was suspected on the grounds of a single feather in a hat seen in 1913, but the whole bird wasn't found (in a museum in Belgium, where it had been misidentified) until 1936. (To understand why scientists are so bashful about declaring the existence of new species, read the salutary tales in the Miyako Kingfisher essay.)

Some of the same people who found the bulbul also managed, amazingly, to co-discover another new species at roughly the same time and in more or less the same region. This was the Limestone Leaf Warbler, a much prettier, more conventional-looking songbird with green plumage and a black brow – boringly conventional perhaps, at least for fans of the new bulbul.

If you do want to discover a new bird, there are clearly certain ideal locations. The best tactic would be to go out to a mountain with a few sheer drops in the middle of nowhere and in the middle of the night, in an unstable South American country, and just hope for the best. If by some miracle you survive, you may be rewarded with a new bird.

Another safer way – and it is hard to think of anything *more* safe – is to beaver away in your laboratory analysing DNA. As we have seen in the Rough-Faced Shag essay, the study of DNA is the new way of splitting into different species birds that were previously thought to be the same. So, for example, in 2007 the Solomon Islands Frogmouth was split from the Marbled Frogmouth after what scientists call 'a taxonomic review' – having another think about what is a species and what isn't. A frogmouth, by the way, looks like a rather evil finger puppet made out of a moulding grey sock. I wouldn't want to spend years in close proximity to it, but researchers hoping to find a new species know that if they kiss enough frogmouths their prince will come – metaphorically speaking of course.

MEINERTZHAGEN'S SNOWFINCH

Tall tales and grand ambitions

Snowfinches are rather obliging friends of birdwatchers. You have to travel high into the mountains to seek them, but as long as you make the effort you can usually find these pretty brown-white-and-black birds, which often come to mountain villages to eat at rubbish tips and other promising feeding grounds. Although they are called snowfinches, these seven or so species are actually members of the sparrow family. Like their cousin, the more familiar House Sparrow, they are insouciant about the proximity of humans, and can be approached quite closely.

These birds' habits are a sharp contrast to those of the wallcreeper, another bird of the mountains of mainland Europe and Asia whose long curved bill and bright red wings make it look rather like a hummingbird that has stubbornly and perversely forsaken the lush forests of the Americas. By contrast with snowfinches, the beautiful wallcreeper is so hard to see that when one incongruously turned up in Paris one winter – the season when they sometimes travel lower in search of food – many British birdwatchers went across the Channel to see it. *Sacré bleu.*

The wallcreeper does however have an Alpine cousin among the snowfinches – the White-Winged Snowfinch, which regularly ventures down to the cosseted civilisation of chalet complexes where it can be admired by humans as a rather unusual part of the après-ski entertainment.

But the discovery of Meinertzhagen's Snowfinch, which prefers the mountains of Afghanistan, right on a particularly sharp and perilous edge of the former British Empire, required a man of stern stuff – a man of the steely timbre of Colonel Richard Meinertzhagen, who described the eponymous bird to science in 1937.

Colonel Meinertzhagen sounds to a tee like one of those worthy but rather pompous characters after whom some of the world's more remote and obscure birds were named in the nineteenth and early twentieth centuries. Born in 1878 into a distinguished and wealthy Anglo-German family – like the Windsors, but posher – he was educated at Harrow, the alma mater of Churchill and various members of the Jordanian and Thai royal families. He then spent his life fighting and spying for his country and exploring for the greater good of human knowledge. Colonel Richard Meinertzhagen CBE DSO finally passed away in his ninetieth year, doubtless happy in the thought that, in addition to the various gongs awarded to him for discoveries vegetable, animal and mineral, his name lived on in the Latin name of the Giant African Forest Hog: *Hylochoerus meinertzhageni*.

But there was – as the Colonel himself might say with the extreme understatement common to men of his background – something a tad odd about the chap.

Meinertzhagen gives us an insight into a seamier side of British society at the time – its habit of turning a blind eye to rogues as long as they were rogues 'from the right sort of people', in other words from the upper echelons of the class system. Just as public-school-educated Anthony Blunt, the eminent art historian exposed as a Soviet spy, was allowed to keep his establishment position for years after the authorities knew about his treasonable activities, Meinertzhagen was a rascal whose constant falsification of bird observations and other wrongdoings were for the most part swept under the carpet until after he died.

One ornithological example, not exposed until careful research in the 1990s, was the discovery that Meinertzhagen had stolen and then relabelled specimens of the redpoll from the Natural History Museum, in an attempt to prove his pet theory that there was a separate British race of this pretty brown-and-red finch, named *Carduelis flammea britannica* by Meinertzhagen (for another famous case of ornithological scandal, read about the Hastings Rarities Affair in the Rüppell's Warbler essay).

There is also, intriguingly, something slightly dodgy about the naming of his snowfinch. It is a golden rule that discoverers of a species never call it after themselves – after finding the creature they must let other grateful scientists do this for them, after the appropriate initial gestures of modest refusal. The Colonel technically named it after his niece, Teresa Meinertzhagen, which in effect allowed him to name it after himself. Calling birds after relatives who share your name is a rare ruse but has been tried a few times. To this vice we owe the splendid names of Mrs Moreau's Warbler and Mrs Hume's Pheasant. But was Teresa actually Richard's niece? People cannot even agree on whether she was in reality a niece, a cousin or a girlfriend whom Meinertzhagen decided to reclassify as a relative to stop tongues wagging. So there is even disagreement about whether she was really Teresa Meinertzhagen, or actually Teresa Clay. You never knew quite where you were with the man, but at the time his exalted background allowed him to get away with murder (possibly quite literally, since his second wife died in a mysterious shooting accident to which he was the only witness).

It is hard to give a conclusive account of any aspect of Meinertzhagen's life because of one very odd character flaw: he was a compulsive liar. His funniest story – that on meeting Hitler he provoked a forty-minute rant by responding to 'Heil Hitler' with 'Heil Meinertzhagen' – is almost certainly untrue, though to be fair to Meinertzhagen, most people's funniest stories are untrue. Meinertzhagen himself acknowledged that he had an 'evil' facet to his personality, and blamed it on sadistic treatment meted out at boarding school.

Did Meinertzhagen's rather unusually obsessive pursuit of birds around the world also stem from his traumatic experiences? One possible motive for choosing an interest that allows you to spend days in the field away from *Homo sapiens* and in the company of other species is that you don't really like *Homo sapiens*. Meinertzhagen's era certainly produced a cadre of upper-class men in the top ranks of the ornithological world, educated at tough boarding schools whose staff thought it was their duty to be nasty to their pupils. Nowadays boarding schools try to be nice to their charges, though it may be mere coincidence that they have stopped producing elite ornithologists. Who knows what shaped Meinertzhagen? If it was a desire to be remembered, he succeeded, partly through the snowfinch's name but most of all through leading such a fascinatingly flawed and fabricated life.

CHOCO VIREO

What's in a name?

Is it right or wrong to sell a bird's name to a wealthy man?

When a species of vireo first described to science in 1996 was named after Bernard Master (Latin name: *Vireo masteri*), eyebrows were raised. Where would it end? Would we have a McDonald's Warbler one day? It seemed to sceptics to be a case of Mammon controlling science.

Master, a retired doctor from the small Ohio town of Worthington, had played no part in finding the bird, and did not see it until many years afterwards. He just gave $70,000 to conservation and in return was commemorated forever (or for as long as this endangered species lasts) in its name.

The bird's non-scientific name is Choco Vireo after the Choco region of Colombia in whose forest this small insect-eating species with rather a pleasant warble was found. It may seem strange that it took so long for it to be discovered, but it lives in a notoriously inhospitable region. The climate is wet and insect-ridden at the best of times. At the worst of times – and that's most of the year – fog and mist make observation very difficult. On top of all this, the vireo lives at the tops of trees, in the canopy of the forest where it is very hard to see. The bird was first found in 1991. But it died soon after it was trapped in a net, and was then rendered unrecognisable after being half-eaten by ants – encapsulating what treacherous terrain this is. A second attempt at procuring a specimen a year later proved more successful.

Having identified the bird, let's mull over this moral conundrum of its name from the bird's point of view.

Master's money allowed conservationists to buy the land where it was first found and set up a nature reserve there, creating at least one place where the bird could be a lot safer. Even the controversy over the naming was good for it, by raising awareness. Remember the adage that no publicity is bad publicity.

Was the step unprecedented anyway, or merely a return to the past? In Victorian times it was common to name the birds you had found after your sponsors. Edward Stanley, the thirteenth Earl of Derby and a politician who had an interest in nature but lacked the time or inclination to go looking for new specimens himself, financed many expeditions. As a result, no fewer than four species are named after him: Stanley's Parakeet, Crane, Bustard and Thornbill.

Lord Derby was highly knowledgeable about birds, and so is Bernard Master – as a world lister (someone who sees as many species of bird as possible by travelling the world) he has visited more than eighty countries to pursue his passion for them. He did not, in fact, see the bird named after him until fourteen years after it was described to science, but this was largely because it was discovered in an area beset by civil war. His trip to see it – he finally spotted it on New Year's Day 2010 – brought yet another benefit. It was discovered in a new location by his tour guide, because the guide was trying to find a different and less dangerous place to show it to Master.

The survival of many bird species will, surely, hinge on the goodwill of individual rich enthusiasts such as Master – just as it always has. Why doesn't a rich benefactor buy the wintering grounds of the endangered Spoon-Billed Sandpiper in Myanmar, to stop it from being hunted to extinction?

Meanwhile, *Vireo masteri* could have done much worse. Master is a pretty grand surname as surnames go. The bird's sugar daddy could just as easily have been Doctor Pratt.

RÜPPELL'S WARBLER

Skulduggery among the bushes

Rüppell's Warbler is named after a nineteenth-century German naturalist, Eduard Rüppell, who typifies the adage that the best way to make a small fortune is to start off with a large one. The son of a wealthy banker, he spent much of his inherited fortune on expeditions around the Mediterranean and North Africa. Rüppell's reward was to have a huge number of species of all kinds of creature, including one fox, three bats and eight birds, named after him, though many, like the warbler whose breeding grounds are in the eastern Mediterranean, were found by others but named in his honour.

Rüppell's Warbler is a handsome but rather severe-looking bird, whose white downward-sloping moustache (technically called a malar stripe) on a black face gives it the air of a stern and strait-laced Victorian gentleman.

But when not one but two specimens of this particular little gentleman turned up near the Sussex town of Hastings in 1914, had the good old-fashioned Victorian value of honest dealing been breached in favour of a Rüppell rip-off?

The Hastings Rarities Affair is one of ornithology's most dastardly cases of fraud – of birds sold at high prices on the pretence that they had been found in Britain. But it was a strange case, since the chicanery was proved beyond all reasonable doubt not by direct evidence, not by some dramatic deathbed confession, but by mathematics.

In 1892 a narrow patch of southern England – a 20-mile radius around the Sussex town – began to enjoy an unusually purple patch of sightings (or more precisely shootings, since this was how birds were collected in those days). Most of these were reported by a man named George Bristow, the local taxidermist in the small town of St Leonards-on-Sea. These included many birds new to Britain, such as the seabird Cory's Shearwater, the wader Terek Sandpiper, and the White-Winged Snowfinch of the Alps – as well as the Rüppell's Warbler. Selling these rare specimens made Bristow a rich man.

'So what?' you might say. Bristow never claimed the feat of having collected all these birds single-handedly – he had a small army of shooters and collectors who did that for him. He was, in a way, like a present-day county recorder – the local person who gathers together all the records.

What's more, there are small patches of land that, these days, seem to hold a magnetic attraction for rare birds, such as the Isles of Scilly off Cornwall, and Fair Isle, an isolated spot far north of the Scottish mainland. Why not Hastings too?

But quite quickly, people began to harbour suspicions about Bristow's activities – to the point that early in Bristow's career, one leading ornithologist described his records as 'fishy'. It was not until 1962, however – more than forty years after Bristow had stopped submitting these records after some gentle nodding and winking from the ornithological establishment – that they were finally expunged from the British list.

Bristow was a clever rogue – but like all clever rogues in cop dramas, he found it impossible to commit the absolutely perfect crime. A few clues – tiny, incidental things – revealed his fraud…

- Clue number one: none of the records are of big birds. That's distinctly odd, because big birds should be easier to spot and shoot. Surely there weren't any because Bristow knew that people would start saying, 'Why did no one else spot them as well?'

- Clue number two: many of the records are of extremely rare birds in groups – a phenomenon that happens very rarely in birdwatching. Usually vagrants are just that – single individuals that were supposed to migrate to Africa, for example, but veered off their path.

- Clue number three: a high proportion of the records, including the Rüppell's Warbler, had graced the bushes and fields of one particular inland village, Westfield, where Bristow knew a local who shot birds. Why would so many rare birds end up in Westfield? Sightings of accidentals are always concentrated on the coast, where they've first reached land after spending hundreds or thousands of miles going in the wrong direction – possibly with a mixture of relief at being able to rest and that sinking feeling that, since none of their colleagues are there, they may just be in the wrong place.

- Clue number four: Bristow usually gave suspiciously common surnames for the identity of his shooters, when questioned by people who tried to follow up records. But in a letter to a leading naturalist protesting the rumours of skulduggery, he referred to a rare kite – a bird of prey – given to him by someone called Glyde. This is suspiciously similar to Glede, the old name for kite – a rather obvious case of Bristow deciding on the bird and then deciding on the name.

But it was statistics, rather than these small giveaways, that finally dealt the death blow to Bristow's credibility fifteen years after his own demise. In 1962 two pillars of the bird world, Max Nicholson and James Ferguson-Lees, used maths to explain just how improbable to the point of fantastical it was that Bristow's records could be true. The lynchpin of their argument was that if you compared the number of rarities in the Hastings area between 1895 and

1924 (when they were submitted), both with other records in other areas, and with records from the same area in modern times, such a concentration was statistically impossible. The records were rejected, which meant that a huge tally of twenty-nine species or subspecies initially recorded as firsts for Britain were struck off.

Where did Bristow manage to get these birds, if they didn't come here through old-fashioned wing power? It seems that he had them smuggled in from abroad, refrigerated to keep them in good condition so that they would look like freshly shot birds. The national newspapers in 1970 found a ship's steward who said he had personally witnessed this – although defenders of Bristow could argue the steward was lying.

Conspiracy theorists might fear, or perhaps hope, that modern juicy scandals of the scale of the Hastings Rarities Affair are waiting to be unveiled, but like most conspiracy theorists they will die disappointed (though if you have a liking for scandal, you will already have enjoyed reading about Richard Meinertzhagen's murky activities in the Meinertzhagen's Snowfinch essay). It was very much a crime of its time, which could only be committed at that particular point in history before there was blanket coverage of Britain's coasts by birdwatchers, many armed with cameras. No one would dare claiming fistfuls of uncorroborated firsts for Britain now – not at least without photographic evidence.

In a strange kind of way, though, Bristow has had the last laugh. Virtually all of his fake firsts are now back on the British list because of other records – in fact many of them had already been seen by bona fide observers before 1962, though after Bristow had first reported them. One, the Cory's Shearwater, is even spotted in large numbers at times, and another, the Cetti's Warbler, as we have seen, now breeds in Britain. The Rüppell's Warbler only had to wait fifteen years before a valid sighting was made. The White-Winged Snowfinch is one of the few birds that hasn't turned up.

Defenders of Bristow have used the belated appearance of so many of Bristow's birds to suggest he wasn't a fraudster after all. I don't think we can go as far as this, but we can at least credit him as a pretty good forecaster of what birds might feasibly rear their heads in Britain at some time. He was a lot better at it than many of the present-day birdwatchers who play similar guessing games in the pub – ever ready, of course, to down their pints to chase any reports of rare Rüppell's Warblers.

STARLING

A bird that needs no introduction

The starling (more properly known as the European Starling), a glossy bird that looks black, green or purple depending on the whims of the sun and the angle from which you're looking, has been a familiar bird in the Old World for centuries. Mozart even kept a pet starling – possibly flattered into procuring it by the fact that this particular mimic was already a bit of a musical star and could sing part of his Piano Concerto No. 17 in G Major. Who wouldn't fall for a bird that can whistle your tune?

But now the starling can also be seen throughout the United States. How did this come about? It is a strange tale indeed, worthy of one of Mozart's comic operas.

Taking a much-loved or highly useful bird or animal and putting it in a place where, by some accident of geography, it doesn't happen to be, seems at first sight to be an eminently sober and sensible idea.

By 1877 New York chemist Eugene Schieffelin had become chairman of the American Acclimatization Society, which rejoiced in the noble aim of introducing 'such foreign varieties of the animal and vegetable kingdom as may be useful or interesting'.

Schieffelin liked the idea of propagating those species of bird that settlers to the United States had enjoyed in their own lands, if they were pretty, or charming, or had a cultural significance which settlers could appreciate. Some have said that, in line with this logic, his ultimate ambition was to introduce every bird mentioned by Shakespeare – of which there are a lot, since he found scores of useful metaphors in them. In fact Shakespeare's frequent reference, in *The Taming of the Shrew* and elsewhere, to a wayward woman as like a 'haggard' hawk – a trained bird that disobeys its master by flying off – has been used by those conspiracy theorists who think he was not the real author of his plays. Some have employed it to press the claim that he was really the Earl of Oxford, who used the same metaphor in precisely the same way in his work.

Schieffelin found powerful friends to back him in his bid to introduce birds to the US. Alfred Edwards, a wealthy New York silk merchant and fellow member of the society, sponsored a system of nest boxes around Manhattan

for House Sparrows to breed in, though sceptics might say he could have found better things to do with his money. The distinguished US poet William Cullen Bryant praised these attempts to introduce the sparrow. He even wrote an over-romanticised poem about the arrival of this rather humdrum bird, 'The Olde-World Sparrow', which declared that:

> A winged settler has taken his place
> With Teutons and Men of the Celtic race

Copies of the poem can be purchased in Ye Olde Gifte Shoppes.

However, Schieffelin was not even the first man in New York to have the idea of introducing new species, since in the 1860s the commissioners of Central Park had already released House Sparrows, chaffinches and blackbirds. This was perfectly commonplace behaviour.

Such thoughts had an even earlier pedigree. In 1849 the French naturalist Isidore Geoffroy Saint-Hilaire called on his government to introduce foreign animals to serve as meat and pest controllers. He had founded the original Acclimatisation Society, which had inspired the US counterpart, and other such groups among European settlers across the world.

But history has made horribly clear that it was all a colossal mistake. Attractive British birds introduced by Schieffelin such as the skylark and bullfinch failed to thrive. But possibly the most successful bird was the least attractive one: the starling. New York City is a noisy place for many reasons: the traffic, the crowds, the air conditioning vaults, Howard Stern, and, last but not least, the starlings, which keep up a constant and varied chatter with a rather electronic timbre to it, like the irritating sound of someone passing through all the stations on the radio but never getting to the right one. One assumes these expert mimics have learnt to say 'Have a nice day!' with that idiosyncratic New York upward lilt.

Having taken a bite out of the Big Apple, starlings have since spread throughout the United States – aided by their versatile bills, which enable them to eat a wide variety of food from caterpillars to fruit. There are now about 200 million of them across North America – all descended from the hundred or so that Schieffelin introduced into Central Park. Far from being a harmless creature, the starling has displaced from their homes native American birds which also like nesting in cavities, such as the Purple Martin and Wood Duck. But why did he introduce the starling in the first place? One can understand the aesthetic appeal behind the society's introduction of larks, robins and tits, which according to the society 'contributed to the beauty of the groves and fields' – although all these pretty birds died out. A possible answer is the starling's brief moment of Shakespearean stardom in *Henry IV Part One*, where the Swan of Avon mentions its extraordinary ability for mimicry.

Naturalists now refer to Schieffelin as a 'lunatic', or even 'infamous'. But hindsight is a wonderful thing. In ornithology, as in all other areas of knowledge, what was done in the past may seem ridiculous and even sinister to us now. But it might well have appeared perfectly harmless at the time. By the late nineteenth century we certainly knew that introducing cats and rats onto islands was a bad idea for conservation, but we didn't really understand this for birds. We have learnt a lot since then, but our knowledge has come a bit late. It's notoriously difficult to exterminate an introduced species once it has spread – closing the birdcage door after the bird has bolted.

RED-BILLED QUELEA

The commonest bird in the world

They call it the locust bird, and when you see the flocks you can understand why.

This creature, only about the size of a chiffchaff, looks harmless enough on its own. But it is rarely on its own, and that is precisely the problem. There are at least one-and-a-half billion Red-Billed Queleas squeezed into sub-Saharan Africa – a greater population than any other wild bird in the world – and they move in huge flocks around the continent that can reach two million strong. That's like having all the chiffchaffs that breed in Britain and Ireland together in the same place. Their stout, strong bills are used to eat the seeds of crops such as wheat and millet, so they can devastate a farm's fields in minutes.

The danger they pose to livelihoods is increased by their unpredictability. Red-Billed Queleas are nomadic. They breed in huge colonies where recent heavy rainfall has made wild grasses abundant. But these queleas will eat the local crops as well as the grasses – destroying an area's livelihood within weeks. They are voracious: a bird consumes about half its body weight a day, equal to about ten grammes.

Ancient accounts show that Red-Billed Queleas were pests back even in Ancient Egyptian times, but the problem has grown worse in recent years. Intensive farming has had the effect of creating factory farms for the birds since the 1970s, allowing their numbers to increase in some parts of Africa by at least ten times in just a few decades, since they breed up to three times a year – if food is abundant, their population can increase very rapidly. Other birds such as the dickcissel of the Americas can be a nuisance to farmers, but it is the combination of the Red-Billed Quelea's superabundance and its unpredictability that makes it a pest in a league of its own among birds. Queleas do not follow normal seasonal migrations from north to south or east to west, but instead move to where the food is.

Scientists are trying to help governments through a pan-African early warning system that forecasts where the birds might go by looking at recent rainfall. There is usually a short time lag between the rain falling and the queleas arriving – so, in theory, perfect organisation should make it a lot easier for governments to predict their enemy's movements. Every week, an Internet map of Africa shows the most likely places for the queleas to attack next.

What can people do if they know the queleas are coming? They can make preparations for extermination. Flamethrowers fired on their roosting sites – so densely packed that the combined weight of queleas can break the branches that they're resting on – have enjoyed partial success, but dynamiting their nesting colonies has been most effective. Planes can also spray crops with a pesticide colloquially known as quelea-tox. But most poor African farmers don't have the money for these sophisticated techniques, so they have to rely on ancient methods that are highly labour-intensive. For many the battle against the quelea comes down in the end to a man standing in each field scaring off the birds by shouting at them. Queleas cause agricultural damage of more than $50m a year, and much of this is to low-income farmers who can ill afford it.

Scientists estimate that farmers manage to kill more than 50 million Red-Billed Queleas a year. That seems a huge number – far greater than the population of even our commonest birds in Britain. But it cannot stem the proliferation of the Red-Billed, a fast-breeding bird which has thrived by taking advantage of the perfect cocktail of ingredients created by humanity.

However, a cautionary note is needed. Consider the case of the Carolina Parakeet, a beautiful multicoloured parrot-like species found in the United States that was common enough to be considered a pest by farmers, to whose fruit it was partial. Since the large flocks which the species favoured used to gather round dead or injured specimens, it was relatively easy to exterminate the flocks by killing one bird and then carrying on shooting. The Carolina Parakeet became extinct in 1918 after the last surviving specimen died in the Cincinnati Zoo, in the same cage in which the world's last Passenger Pigeon had expired four years before – making it a Death Row for bird species.

Occasionally birds can be pests, and their numbers need to be controlled. But if we do find a way of killing Red-Billeds effectively, we should remember that there comes a point when the killing has to end.

CARRION CROW

Birdbrains

In Japan some Carrion Crows have developed an ingenious technique for preparing their food. They get humans to do it for them. The crows place walnuts on the road – when the traffic lights are red of course, so that they won't get run over. They then retrieve them after the cars have broken the hard shells by driving onto them. Britain has Carrion Crows too, but they haven't yet thought up this wheeze – though an amateur observer in Edinburgh has seen them dunk biscuits in water, using the liquid in exactly the same way that builders are universally reputed to use tea.

Are these particular Japanese Carrion Crows the cleverest birds in the world? Many scientists would say the New Caledonian Crows of the Pacific are even brighter. They use their beaks to make hooks out of twigs, and then employ the hooks to pull insects out of holes.

It also appears that cormorants can count. A scientist has observed fishermen in China using cormorants to go fishing for them – but noticed that the birds will only do so if bribed with every eighth fish. If they don't get the eighth fish, rather than the ninth or tenth, they will go on strike, sitting sullenly on their posts.

These are all examples of a limited, mechanical intelligence that kicks in when food is involved. But scientists have also noticed birds' social intelligence – an acute knowledge of where they stand in the hierarchy and how this must shape their behaviour. Konrad Lorenz, the Austrian scientist we have already come across in the Herring Gull and Song Thrush essays, wrote in the 1950s about the complex goings-on among a group of tame jackdaws that he kept. These members of the crow family are common in Britain too – smaller than Carrion Crows and with grey hoods at the back of their necks, but otherwise similar-looking. The jackdaws had a very strict pecking order. Sometimes this was quite literal – if necessary, a senior bird stabbed a junior one with its beak in order to maintain the required hierarchy – but for the most part physical attack wasn't necessary because each bird knew its position in the group. So the top birds occasionally pecked the birds that were next in line, to remind them who was boss, but they didn't generally bother with the jackdaws in the group that were further down than that, because they didn't need to. These even more junior jackdaws knew their place. They were rewarded for this by being ignored by the birds at the top – which was the best they could hope for.

However, the birds somewhere in the middle in seniority terms still menaced these birds at the bottom. If this seems immensely complex it became even more so when the top male jackdaw took a fancy to a very junior female jackdaw, like a Victorian aristocrat falling for a match girl. Once the female jackdaw had realised this, she took advantage of her place in the hierarchy to maltreat those birds that were now below her – showing her ability to understand her rapidly altered social status, as well as a shocking ruthlessness that would put a social climber in a Jane Austen novel to shame.

Not all birds are as clever as this, though. Crows can generally solve 'the detour test' – understanding the concept that they have to move further away from the food they want in order to get it. The trial, done by putting a glass barrier between the bird and the food, seems pretty simple for crows that can get humans to crack their own nuts for them. But hens fail this same test – so they really are pretty dim. Hens are also not clever enough to realise that maintaining a pecking order doesn't require much pecking. So, unlike jackdaws, they go to the trouble of troubling every bird below them in status.

But some scientists regard parrots as even cleverer than the crow family. An early contender for parrot genius was the pet macaw of Sir Joshua Reynolds, the eighteenth-century English portrait painter, which regularly appeared in his portraits of human subjects. Who needs the Mona Lisa as artist's muse, when they can have a macaw? It is said that, having taken a dislike to Sir Joshua's maidservant, the macaw attacked a picture of her done by one of his pupils – in one fell swoop from its perch, proving both that it was clever enough to tell people apart, and that its master was an excellent teacher of painting.

In the twenty-first century, two parrots stand out as veritable avian Einsteins, both of them African Greys. Alex (short for Avian Learning Experiment) was able to identify fifty different objects – describing the shape, colour and material of each one. N'Kisi has a vocabulary of about 1,000 words, and is able to use them to make his own independent intelligent observations. He can string words together, using new combinations that he has not been taught – proving that he isn't just learning mimicry. So, for example, he judges aromatherapy oils as 'pretty smell medicine'. N'Kisi was also, fittingly, once shown pictures of Jane Goodall, the expert on apes who had done so much to prove *their* intelligence, with her beloved chimpanzees. When she visited him one day at his New York home, his greeting was to the point, intelligent, and possibly even deliberately funny: 'Got a chimp?'

YELLOWHAMMER

A cheese sandwich without the cheese

There comes a point in every nature lover's existence where they ponder one of the great questions of life: 'Can I really be bothered to learn birdsong?'

It seems so much harder than learning to understand a foreign language, the meter of poetry, or even the complete works of Søren Kierkegaard in the original Danish (though it may not be tougher than deciphering Chaucer). One of the great problems is that birds don't have lips, so they don't make consonant sounds. Bird identification guides have decided to address this knotty problem by pretending it doesn't exist. So they assign random consonants to sounds. A Great Spotted Woodpecker's one-note call is sometimes described as 'kik', sometimes 'chik' and sometimes 'pik', and so on until almost all the consonants have been used up.

One technique is to imagine what a bird might be saying if it was using human language. The best known of these examples in Britain is probably the yellowhammer's 'little bit of bread and no cheese' – singularly appropriate given the non-lactose diet of this fine farmland bunting with a bright yellow head streaked with black, and not a bad rendition of the yellowhammer's song either (and it's such a bizarre sentence that you'll remember it too). The yellowhammer doesn't flit around showily, asking to be seen, like some finches. Instead it just sits on a shrub calling rather softly – motionless except for its opening and closing bill. Often it is at the very top of the bush, so you can get a great view of it if you know where to look – but you probably wouldn't see it if you didn't know its song.

An equally apposite verbal description is the 'go-back, go-back' call of the Red Grouse. Given that people are often approaching to shoot it, this seems a perfectly rational thing to say. There is also the Great Tit's four-note 'teacher! teacher!', which nicely captures its rather insistent quality, and the woodpigeon's 'two cows, Taffy', as if a cast member of *The Archers* is ordering a couple of bovine specimens from a Welshman.

Outside Britain a common aide-memoire is the 'quick, doctor, quick' of the Common Bulbul, a rather plain brown, skylark-sized bird in Africa. The Ivory-Billed Woodpecker of the Americas, now thought to be extinct, made a sound like a gun going off – which led to no end of confusion when an enthusiast declared he had rediscovered it after taping a sound recording that turned out, after all, to be a gunshot. But once abroad, we run into the problem that a bird will of course appear to be saying a different thing in different languages. So in Anglophone Africa, the Red-Eyed Dove is emphasising, in that insistent way that doves call, 'I AM a Red-Eyed Dove', but in Francophone Africa it is saying, 'Je PLEU-re-re-re-re' – 'I'm CRY-i-i-i-ing'.

In reality it doesn't matter what we think a bird is saying, as long as it works for us – but this technique only comes in handy for the minority of birds whose call consists of a limited number of sounds. So we can't use it for the blackcap, for example, which seems to go on forever – though in a very pretty way, like an opera singer who can't resist prolonging his aria with ever more flourishes.

But in reality learning birdsong is not nearly as hard as it seems – provided you're prepared to put in the initial effort. Once you have learnt merely a few songs, you can learn them all – or almost so. If you know the wren, robin and blackbird – all birds that sing in almost any remotely leafy area in Britain, even in towns – that is enough to form the essential base that allows you to learn songs by making comparisons. So, for example, the dunnock's song is like the wren's but sounds less aggressive, the whitethroat's song is like the dunnock's but a bit scratchier, the blackcap's is like the whitethroat's but longer and a bit more flute-like – and so on. Thanks to the MP3 player, you no longer even have to remember these pointers after listening to them on CD at home – many birdwatchers download the calls and songs of hundreds of birds onto these gadgets, for handy reference in the field.

One great advantage of learning birdsong is that you will always hear more birds than you see. On official breeding bird surveys the vast bulk of the birds that are recorded are found through song. If you're just on a casual stroll and you hear a nightingale but it refuses to come out of its bush, you still have the satisfaction of having located it.

One might think that our view of what a bird is saying when it's singing does not alter throughout the centuries, but this, too, is changed by what we are familiar with in different ages. Some of the calls of petrels – mysterious seabirds that fly far out at sea all day, coming to their nesting burrows at night – used to be likened to the laughter of demons, or even, less romantically, to a fairy throwing up after a wild night out with other fairies, but now that people no longer generally believe in such things, both likenesses have lost their descriptive power. People are more prone to liken them these days to crying babies, since we can all testify to the existence of wailing infants. Meanwhile, the call of the Lesser Redpoll, a kind of finch, used to sound exactly like a telephone ringing – before the customised ring tones of mobiles took over the world and made this description obsolete. A redpoll certainly doesn't sound like Beyoncé's latest hit, and both would probably be insulted at the suggestion.

Has the restrained message of the yellowhammer's song become obsolete too, in an era when we are told that we are facing an obesity epidemic? How about an update to 'little bit of bread and no cheese'? 'Happy Meal, Cola and French fries'?

AZURE-WINGED MAGPIE

What are you doing here?

We can understand what has made ornithological explorers roam the world's most unwelcoming and downright dangerous places, overcoming snowstorms and sandstorms, frostbite and malaria (and for a tale of *sans-pareil* suffering in search of birds, read the Emperor Penguin essay). What drove them on was the prospect that they would discover a bird new to science – ideally one of resplendent plumage and sleek proportions.

So think of those poor explorers who endured all the usual trials and tribulations that come with the job, only to discover a bird thousands of miles away that looks like one in their own backyard.

A perfect example of this is the Azure-Winged Magpie of eastern Asia. These large but graceful birds with blue wings, an extremely long blue tail and a smart black cap, are constantly passing overhead on any trip to Beijing's majestic Summer Palace, surrounded by the verdant woodland that forms the magpies' habitat. They are common even in quite urban areas of Tokyo, and have won the dubious honour of being named the official bird of Setagaya, one of the city's twenty-three wards – which is a bit like being the borough bird of Croydon.

It is one of the oddities of ornithology that they are also found in Spain and Portugal at the tip of western Europe – but nowhere else in between. The two populations are separated from each other by the rest of Europe and most of Asia. One possible explanation is that Iberian sailors took a fancy to these beautiful birds, and brought them back home as souvenirs, in the days before Hard Rock Cafe T-shirts existed.

The world is full of such avian anomalies – widely separated populations of the same bird. There are Blue Tits in the northern highlands of Jordan. You might be underwhelmed by this fact, since these birds are in your garden – but this group is about 300 miles away from the nearest gathering of other Blue Tits. We have a similar case in Britain too. The prettiest of Britain's members of the tit family, the Crested Tit, has a black-and-white crown that tapers to a fine plume at the top. In the United Kingdom it is found only in the pine forests of the Highlands of Scotland, hundreds of miles away from northern France and southern Norway, its nearest strongholds.

If you don't believe the jolly tar souvenir hypothesis, there are other explanations of how this phenomenon might happen. The most likely, in most cases, is that these birds form 'relict' populations. That means, for example, that the Azure-Winged Magpie used to be found right the way across the whole of Europe and Asia, and that the populations that remain are therefore the relicts of an earlier age.

But for some birds there is another, opposite explanation – that isolated populations establish themselves by taking a bold leap into the unknown that pays off. Take the Two-Barred Crossbill, which like other crossbill species has an upper and lower bill that literally cross at the end. They use them like pincers to take out the seeds from the centre of pine cones. If I tell you that one of the populations is in Alaska, Canada and the northern United States, and another is in northern Asia and north-eastern Europe, can you guess where the third population is? Whatever your guess is, it's almost certain to be wrong, since the answer is Hispaniola, an island nestling in the middle of the sunny Caribbean.

In the Two-Barred Crossbill's case, it is possible that all the populations between Hispaniola and the northern part of North America died out. But it is also possible that some crossbills far out of their usual range landed there and liked it.

This is because crossbills are 'irruptive' birds – like many species of the far north, including Snowy Owls and waxwings (exotic-looking multicoloured birds with touches of red, yellow, white, brown and black that visit Britain in the winter). This is the ornithological equivalent of the conundrum that for a long time you don't get any buses, but then three of them come at once. Irruptive species are usually content to stay in their normal territory, but in years when the food supply runs low, they spread out to new territory in large numbers.

As usual, humanity complicates matters. There has been much hand-wringing over the mysterious find of a Two-Barred Crossbill corpse in 2007 in Florida's Everglades National Park. Did it fly all the way from Canada, or did it come from the Hispaniola population? The most likely explanation is rather unromantic, unfortunately – it was most likely a particularly inattentive bird that was hit by a carload of particularly inattentive Canadian holidaymakers driving down to Florida for their annual holiday, and simply fell off when they reached their destination. How's that for a vacation?

The twist in this tale is that those scientists who hundreds of years ago discovered these far distant versions of the birds that were in their own back garden are belatedly being helped out by scientific breakthroughs. The trend for splitting birds into different species now means that Hispaniola's Two-Barred Crossbills are classed as their own species, the rather unimaginatively named Hispaniolan Crossbill. A recent paper has also suggested that the Azure-Winged Magpie should be split into separate Asian and European species. It appears that in both cases the birds' different populations have been separated long enough for them to evolve into distinct – though extremely similar – species. The only problem is that in many cases, no one can remember who found such belatedly new species in the first place (because, of course, they weren't thought to be new species). So the discoverers are unlikely to have their names bestowed on these new birds. Such are the cruelties of history.

PALLAS'S WARBLER

Opera, war and birds

When birdwatchers are scouring bushes at some windy spot on Britain's east coast on a cold morning in late autumn, they're looking for a bird that will prevent their suffering from having been in vain. Finding an all-too-common wren buried in the bushes is not regarded as fair recompense for the privation of a chilled birder – and if some sense of moral justice is seeping through to the reader like the chill of a Norfolk October, that is deliberate. Birdwatchers feel a sense of metaphysical outrage when a morning's adversity is not rewarded by a bird even slightly out of the ordinary.

The bird in many of their fantasies – for birders have fantasies about seeing a particular species, just as football fans have about their team winning the FA Cup, and teenage boys have about girls – is the rare Pallas's Warbler.

It is a tiny bird even by the standards of tiny birds, at only 9 centimetres long from head to tail, which makes it smaller even than a Blue Tit. But this green warbler manages to fit an enormous number of fetching yellow bands within its small frame – a thick yellow stripe across the rump, two on each wing, one at the top of the head and one just above each eye. It is, in short, like a tiny flying tiger that has taken on the colour of the vegetation among which it roams.

The Pallas's Warbler had never been seen in Britain, until one was shot while trying to mind its own business flitting about the long grass in 1896 at Cley-next-the-Sea, the Norfolk village celebrated for attracting attractive birds. The next one was not found until 1951, and this extreme rarity seems eminently logical, since it is supposed to breed thousands of miles away in Siberia and winter in warmer parts of Asia.

But in recent decades Pallas's Warblers have been falling from the skies – and I use the expression only semi-metaphorically. Currently about forty are seen here each year, usually in the autumn. When you consider, in the age of mass email and text communication, how many people receive news of each sighting, it's clear that if you really wanted to see a Pallas's Warbler in any particular year in Britain, you could probably do so.

The most obvious logical explanations for this bounty are either that the Pallas's Warbler's global numbers have multiplied, or that its breeding

range has expanded significantly westwards, bringing it nearer Britain. Most inconveniently, neither of these points is clearly true, so to understand the reason, we need to broaden our minds beyond everyday logic to immerse ourselves in esoteric Eastern philosophy.

It is sometimes asked in Zen teaching: if a tree falls in the forest and there is no one there to hear it, does it make a sound? If a Pallas's Warbler lands on a bush in Norfolk and no birdwatchers are there to train their modern, high-tech, and surprisingly light binoculars on it, did it come to Britain at all?

The probability is that Pallas's Warblers have been reaching Britain for centuries, but we never noticed. The fact that we did not do so, even though the isle's inhabitants began to show a distinct interest in birds from the eighteenth century onwards, is testament not to our stupidity, but to our lack of the right optical equipment.

It was our love of opera that kick-started the historical advance towards modern binoculars, then our love of war, and only finally our love of birds. The first binoculars that were better than useless were opera glasses, but the early ones tended to magnify objects only by a factor of three or so, and were not much good for looking at objects such as one's fellow opera-goers off the well-lit stage. But the military soon realised that what could be used to watch a soprano prancing about in an opera house could also be used to spy enemy soldiers stealthily making their way across fields, and this gave new impetus to innovation in binoculars. However, having peered through binoculars first used by a military officer in World War Two, I can testify that even by this stage they left a lot to be desired. It was only the explosion in popularity of birdwatching that brought binoculars up to the high present-day standard, where strong magnification was matched, for the first time, by technological advances that maximised the amount of light suffusing the view of the bird – turning it from a dark silhouette into a creature of subtle and finely coloured plumage. This allowed a Pallas's Warbler to be identified in seconds, which is all you'll get since they are notorious for never staying on the same bush for very long.

But the belated realisation that Pallas's Warblers regularly come to Britain creates a brand-new intellectual problem: why do they do this, since most of their comrades are half a world away?

There are two explanations for why so many travel here. One is that they are very stupid, and the other is that they are very clever.

It is possible that those which come to Britain are correctly imbued with their evolutionary urge to migrate, but not with that essential final touch – the urge to migrate in the right direction. In other words, they may be aberrant individuals who go the wrong way. If they accidentally head in the wrong direction when they start their migration – going a few thousand miles to the west instead of a few thousand miles to the south-east – many will end up in Britain. So in the survival of the fittest, they are the least fit. But it doesn't matter for the species as a whole, because the bulk of them still go the right way.

But it is also possible that those who come to Britain are scouting parties – birds that have come here to see if there are suitable wintering sites. In this version of events, they are like the hardy pioneers in westerns. 'Wagons roll!' as homesteaders are prone to declaiming in the movies – though the pioneering Pallas's Warblers are less gung-ho. They prefer a decidedly more downbeat one-note 'sweeep', a plaintive call that sometimes (and more now than ever before) draws our attention to this little green-and-yellow gem's miraculous presence in Britain.

DARTFORD WARBLER

A beneficiary of climate change?

The Dartford Warbler is one of the few birds in the world named after a British place name – the spot where it was first found in 1773 – but this seems a strange country for it to be in at all.

The warbler has always held a precarious clawhold in England, because of its stubborn refusal to migrate in the winter to warmer climes where there will be plenty of insects. After the harsh winter of 1962–3 it was reduced to a hardy band of less than a dozen pairs, like the tiny number of French soldiers who survived Napoleon's winter retreat from Moscow. '*Vive la Fauvette Pitchou*', as they say in France (or should say, if they are not already doing so). Numbers have recovered to about 3,000 pairs since then, but any harsh winter devastates the population once more. It is, in short, very prone to what Gordon Brown used to call 'the cycle of boom and bust', before boom turned to bust and he had to change his rhetoric, shortly followed by his address.

This striking reddish-purple warbler has white spots on the throat that look like a beard and give it the impression of an old man – but it seems a stubborn old sort of fellow rather than a wise one. It has managed to survive in Britain because it seems able to recover from population crashes, through reproducing at a very rapid rate. But the country has always seemed to be at the outermost limit of its range, with most of the birds breeding in mainland Europe. However, scientists say this precarious existence in Britain may be about to change for the better, because it could be a beneficiary of man-made global warming.

The concept that global warming may be good for any creature sounds heretical, but in the case of Britain's Dartford Warblers it may just be true. If temperatures rise we will have less harsh winters, and the warbler will stop suffering near-extinctions in our country. Britain could even become a stronghold if places like southern Spain, where it is populous, become too hot and dry for its tastes – producing an arid landscape that supports fewer insects. Our only other warbler that does not migrate – the Cetti's – may also benefit, since it, too, is blighted by similar population crashes during severe winters.

Another global-warming theory is that resident British birds that don't migrate, like the blackbird and House Sparrow, may benefit from climate change too. If winters are milder, then spring will come sooner, and they will be able to start breeding more quickly.

But although I give these two examples of how climate change might be good for a bird – in Britain at least – other birds will inevitably lose out.

This is partly a question of competition. If resident birds start breeding earlier, species such as swallows that migrate to here will be mere runners-up, because the best nest sites will have already been taken, for example.

Overall, scientists have a very hazy idea of how global warming will affect birds, because they don't really know how precisely it will affect the climates of individual countries. The most common theory of the effects of climate change on Britain is that it will make the country grow warmer. But another theory is that global warming could, through a complex mechanism, disrupt the Gulf Stream, the ocean current that keeps our winters so unusually warm for so northerly a country. That could create the opposite effect – Arctic conditions that will exterminate Dartford Warblers but be good for Arctic species such as Snow Buntings. Moreover, scientists say that even if winters are generally warmer, climate change may generate more extreme weather events, including cold snaps – which could be enough to reverse the fortunes of resident warblers.

The key point about climate change is that even if we don't know what it will do, we do know that whatever it does will happen very quickly – almost certainly too quickly for many species. This is because birds will, in most cases, not be able to respond to climate change through the relatively slow process of natural selection (though for an example of a bird that has evolved rapidly, see the Blackcap essay). Huge climate change has already taken place in history, but usually over thousands of years at the very fastest, and this has given many species enough time to react (though others have gone extinct despite the slow pace of change). At the moment swallows migrate at just the right time because they have evolved to do this – and it will take a long time for them to change their habits. Swallows are as stubborn as Dartford Warblers in their own way. Up till now, this has saved them, since they are programmed not to arrive here too early (when it is still winter) or too late (which would reduce the number of days they can spend raising young before they all migrate back to Africa). But in the future swallows' obstinacy could destroy them, just at the point where it causes an upturn in the British fortunes of the Dartford Warbler.

If Britain does grow warmer, it will change in a thousand ways, taking on many of the birds, insects and plant life of mainland Europe. An increase in the Dartford Warbler would simply be one manifestation of this. So in future years Château de Rothschild could be replaced by Château de Romford, and England's Dartford Warblers could even return to Dartford, where they died out last century – making it a topsy-turvy world indeed.

CHIFFCHAFF

A parson's conundrum

It has one of the most simple songs of any British bird, is a uniform greenish-yellow that sometimes verges on grey, and can be frustratingly elusive as it hops about in the middle of bushes. What, then, is the appeal of the chiffchaff, whose name so faithfully sums up its short but rich-throated sound?

Much fuss is made about hearing the first cuckoo of spring, since it is seen as the very harbinger of the season. The first cuckoo record has become a ritual of British life, marked by letters in *The Times* announcing its arrival. But the flaw is that spring invariably arrives a long time before the cuckoo, which is not here till well into April. By contrast, you can hear chiffchaffs on fine sunny days in March. Some even overwinter in Britain these days, though they do not sing till spring.

So the chiffchaff is a more timely bird, but there is more to its appeal than this. It is an exceptionally dainty bird too – one of the three British breeding members of the *Phylloscopus* (literally 'leaf-seeking') group of warblers. All three are perfectly proportioned – their heads, tails and wings seem neither too large nor too small for their bodies, giving them an understated elegance that takes a while to appreciate. They are, in short, the opposite of showy beauties such as the birds of paradise. They also look very much alike, so another attraction of the chiffchaff is the challenge of telling it apart from its two *Phylloscopus* cousins – the Willow Warbler, which is almost a doppelgänger of the bird, and the Wood Warbler.

The fun of identifying birds is one of the greatest joys of birdwatching, and the chiffchaff posed one of the first puzzles of all. Gilbert White, the eighteenth-century Hampshire parson regarded by historians as the world's first birdwatcher, was also the first man to realise that Britain had three different Phylloscs, as they are known for short (though generally by scruffy young birdwatchers rather than parsons). This was a brilliant achievement for three so similar-looking birds, in an age before binoculars. A chiffchaff, by the way, has dark legs, while a Willow Warbler has pale ones. That, apart from their song, is the best way of distinguishing them.

White teased out the truth that the chiffchaff, Willow and Wood Warblers were all different species, but the task of distinguishing other similar British birds from each other continued after White's death when Lieutenant Colonel George Montagu had what we could fairly call the mother of all midlife crises. He managed to cram into his fifth decade an impressively broad and large range of scandals, including an extramarital affair, a court martial, and a litigious dispute with his son that eventually led to the loss of the family fortune. This culminated – most madly of all, in the context of early nineteenth-century society – in his decision to flee to an isolated cottage in Devon to devote the rest of his life to the bizarre pursuit of birdwatching (joined by his scandalous mistress, just to make double-sure the neighbours kept away). Conventional society's loss was birdwatching's gain, since Montagu managed to sort out two very tricky pairs of British birds from each other for the first time. He worked out that the Cirl Bunting – whose English stronghold is still the south Devon countryside in which he lived – was a different bird from the yellowhammer, and that the Montagu's Harrier that bears his name was different from the very similar-looking Hen Harrier. Montagu may have been a lacklustre Lieutenant Colonel, but he was no ordinary ornithologist.

Still, White's fans might argue that his was the greatest triumph of all. At least Montagu enjoyed the advantage that harriers fly around in the open, in their familiar V-shaped flight with wings tilted upwards. But to separate two very small separate species that are usually found deep in foliage must have required the patience of Job, to use an allusion of which the parson would doubtless have approved.

CHATHAM ISLAND ROBIN

Back from the brink

When you are the only eligible female left in the world, your species is as near to extinction as it is possible to be without facing certain doom. But is it possible to come back from so close to the edge? The history of the Chatham Island Robin shows that it is.

By 1980 this robin was down to one fertile female and four other birds on Mangere Island, a speck of land about 500 miles to the east of New Zealand's South Island.

The Short-Tailed Albatross is probably the only bird that possibly came even closer to extinction. Butchered in its millions for feathers, the global tally for this bird that bred in islands off Japan was down to about fifty in 1939. Two years later a volcano covered virtually all of the last island where it bred, Torishima, with lava. The bird was declared extinct.

But in 1950, a small group of the albatrosses was found on Torishima, and they started to breed again. The species had been suspended between eradication and survival for close to a decade. For up to nine years it had, we can only presume, stopped reproducing. Almost any other bird would have died out, but the albatross survived because albatrosses are among the most long-lived of birds, and can put in an innings of more than fifty years.

And the Chatham Island Robin? It was a close-run thing, but the robin managed to survive – with the aid of dedicated and truly inventive human interference. Every year a team of conservationists removed the first clutch of eggs from the one fertile female – and then her descendants – and placed them in the nest of another bird, the tomtit. The female robin relaid her eggs, doubling her productivity. Now there are about 250 robins.

But for many other birds, by this stage intervention is too late.

One reason is the fact that not all people who are interested in birds are interested in birds' welfare. The nineteenth century saw an obsession with collecting birds and their eggs, and the rarer the bird, the greater the prize. By logical extension, the very last clutch of eggs that separated a bird from extinction was the most sought after of all. Even museum curators, who should have known better, joined the chase. The knowledge that the Great Auk – a seabird of the North Atlantic – could soon become extinct, provoked a horde of collectors to descend on its last known colony, off Iceland. The last birds there were killed in 1844, and sent to a museum, where future visitors could tut-tut over the bird's extinction and be thankful at least that museums had preserved specimens for posterity, without knowing their hypocritical role in extirpation. In Victorian times the few remaining eggs of any bird close to oblivion could fetch prices higher than a labourer's annual salary.

And then there is the element of sheer unpredictability, when a species is down to such a small number. A good example is the Stephens Island Wren. Reduced to tiny figures on a tiny island after it was wiped out on mainland New Zealand, it was then finished off either entirely or largely (depending on which account you believe) by the pet cat of the lighthouse keeper. It had taken a liking to hunting these easily caught creatures, which suffered from the unusually bad luck of being one out of only a handful of songbirds known to have been flightless, out of thousands of songbird species. Lucky pussy. Every one of these flightless species, by the way, is now extinct.

But in one sense, the numbers don't matter at all – the crucial determinant is whether human will is for or against you. The return to the hundreds of the Chatham Island Robin shows how quickly birds can multiply when protected. At the other end of the scale is the Passenger Pigeon. Many people think that in the nineteenth century it was the commonest bird in the world, with a population of up to five billion, and one study suggests it accounted for between 25 and 40 per cent of US bird numbers. But it was hunted for meat on an industrial scale precisely because it existed in such a huge quantity, and thereby suffered a very abrupt decline in the twenty years to 1890. The last specimen, Martha, died in Cincinnati Zoo on 1 September 1914, weeks after the pointless slaughter of millions of humans by each other had begun with the opening shots of the Great War. There will be more extinctions, of course – however common a bird species seems now – unless humans prove more prescient. Thankfully, there are signs that conservation organisations are much quicker to jump in when a bird is showing signs of catastrophic decline. The Indian Vulture is a hopeful case in point. Numbers have very rapidly fallen by up to 99 per cent because of a chemical used by vets on cows, which is harmless to the bovine creatures but fatal to the vultures that eat them after they've died – through the same process of biomagnification that once hit the Peregrine (for more on this fearsome hunter, see the Peregrine Falcon essay). Captive breeding programmes have quickly been set up to stop the last 1 per cent or so from disappearing. It may sound too late, but if the Chatham Island Robin was able to survive because of one fertile female, it's hard to be utterly pessimistic. Where there is life, there is hope – and, given the history of Lazarus species, even where there isn't apparent life there is still hope.

SAVI'S WARBLER

The new bird that wasn't so new

The Savi's Warbler, a small reddish-brown bird that does not exactly skulk but certainly minds its own business in English reed beds, was discovered in 1821 by Paolo Savi, an Italian ornithologist and geologist.

But that's not strictly speaking true – or at least it depends on what you mean by 'discovered'. The first specimen was acquired by ornithologists in 1819 in Norfolk – but no one recognised that it was a previously unknown bird. Perhaps we should say it was 'found' in Norfolk in 1819, but then 'discovered' in Italy in 1821.

But is even that true? The men who worked in the Fens had known of the bird for years, but no one had bothered to ask them about it. The bird is similar to the Grasshopper Warbler, which makes a similar sound like a fishing rod being reeled in. The marsh men had for a long time talked of the 'night reeler', as the Savi's Warbler was known to them because it was more prone to sing in the dark than the 'reeler' or Grasshopper Warbler – a separate but rather similar species.

The Savi's Warbler was not the first or last bird 'discovered' by scientists which local people had known about all along. The Calayan Rail from the eponymous Filipino island was first described to science only in 2004, but the islanders had talked about it for years. Often scientists have ignored local reports and ended up with red faces – failing to remember that country people who live in close proximity to nature often notice things that city-dwelling birdwatchers don't, even if they're not equipped with binoculars and other birding paraphernalia. A certain amount of scepticism is still *de rigueur* – local people also talk of still-living Sabre-Toothed Tigers in South America, or of the yeti in the Himalayas, when it is very unlikely that either are there. But open-mindedness is a virtue too. Unfortunately, human beings love creating hierarchies – in rainforest villages, urban offices, and most of all in science – and one of the defining features of a hierarchy is that you don't listen to what people below you say.

We cannot be too hard on the British experts who failed to notice the Savi's Warbler, though, since it is not exactly Britain's most noticeable bird. It is small, brown, rare, and usually buried in reeds. Frustratingly, soon after the warbler was found in Britain it died out – probably because of habitat loss. It only returned a century later, and, as we speak, is on the cusp of disappearing again.

Ornithologists suffered from the other handicap, when locating the warbler, that it is one of those birds that doesn't sound like a bird. The most common descriptions equate it either with fishing rods or insects, but to me it sounds most like electricity pulsing through a telegraph wire. In Britain at least, small warblers hanging about in reeds seem to specialise in unbirdlike sounds – possibly, scientists say, because they are trying to disguise from predators the fact that they are birds. The Sedge and Reed Warblers summon up images of threshing machines that have run out of control and careered into the depths of a reed bed, hotly pursued by a fretful farmworker. These birds are all warblers by name because of their more tuneful cousins like the Willow Warbler, but when it comes to their voices, they are certainly not warblers by nature.

These marshland creatures are not the only birds that do not really sound avian. The Corn Bunting, a fast-declining bird of British farmland, makes a sound like jangling keys. The stonechat emits a quiet sound like someone chopping wood far in the distance. I know this because I once corrected a non-birdwatcher on a Surrey common to point out that we weren't hearing a bird at all, but the sound of a local hewing wood. A stonechat soon popped up at the top of a bush to prove me wrong. Birdwatchers, too, have their hierarchies – which are wont to be overturned.

BLYTH'S REED WARBLER

The madness of crowds

William Eagle Clarke proved, one autumn in 1905, that it was possible for a learned and sufficiently imaginative man to work out where the rare birds were going to be, simply by looking at a map.

Clarke, who worked for the natural history department of a museum in Edinburgh, was trying to make up his mind where to go for his hols. Fascinated by bird migration, he decided to consult a map of Scotland to find a suitable location. His eye fell on Fair Isle, a tiny speck of land measuring three by one-and-a-half miles. Although officially part of Shetland, it lies more than 20 miles from Mainland, the rather misleading name for Shetland's biggest island. It seemed a remote and forbidding place for a vacation, but it was precisely this that attracted Clarke. His reasoning was that there would be plenty of birds there, alighting for rest and food on the only land mass for miles around.

How right he proved to be. Making the pilgrimage there every autumn until 1912, he was rewarded with a cornucopia of rare birds, including a few firsts for Britain. Inspired by his experiences on Fair Isle, Clarke published one of the great ornithological classics, *Studies in Bird Migration*. Fair Isle has retained its reputation ever since as perhaps the best place in the British Isles to find rare birds, and Clarke is remembered by his birding descendants. Without Clarke's logical leap of faith, he would now be forgotten by history and Fair Isle would be known only for its eponymous jumpers – discovered by the great British public in 1921, the year of Clarke's last visit, after the future Edward VIII sported one of these intricately patterned little numbers.

The Blyth's Reed Warbler, a greyish-brown bird that breeds in Russia, was one of the firsts that Clarke detected, in 1910. By this time he had managed to persuade other interested comrades to come with him, including the exalted Duchess of Bedford, one of the few female birdwatchers of the time, who helped him in the chase for the warbler. Identifying the bird was quite a coup, since its close similarity to several similar birds, such as the closely related Reed and Marsh Warblers, makes it an extremely easy bird to overlook.

But there is a twist to the story of the Blyth's Reed Warbler. In the autumn of 1979 another was seen on the Isles of Scilly, which vies with Fair Isle as the chosen destination in Britain for pursuers of rare birds. This particular species is reticent about showing itself at the best of times, but there was an added complication: the bird was on private land, and the landowner decreed that only a few people could be allowed onto it at any one time. Given that hundreds of birdwatchers wanted to see this extremely secretive bird, it presented a daunting challenge – met, of course, by the British genius for queuing. Small groups of birdwatchers were allowed onto the site for a quarter of an hour or so, and had to hope for the very best that the bird, which by temperament does its very best not to be seen, slipped up by making itself briefly visible.

Despite this obstacle, all proceeded remarkably smoothly throughout the day. The bird was regularly glimpsed in the undergrowth, showing off its sober brown plumage, attractive white eye-stripe and reed-coloured legs. This was birdwatcher heaven.

Wait a minute – reed-coloured legs? Screeeech – the sound of mental brakes being applied somewhere in a birdwatcher's supercharged cerebrum. One experienced and highly regarded amateur birder, Peter Grant, pointed out that Blyth's Reed Warblers have dull grey legs, not light brown. Surely that made it a Marsh Warbler – a scarce bird but not in the same legendary league as Blyth's? He was proved right when the bird was caught and closely examined the following day.

So how many birdwatchers had uncritically accepted the Blyth's identification? A few hundred – confirming the enormous power, among human beings, of the magic combination of wishful thinking and the herd mentality. They had been told it was a Blyth's Reed Warbler, they wanted to believe it was a Blyth's Reed Warbler, and they weren't about to question this. Anyone wanting a good example of the madness of crowds need look no further.

WILLOW TIT

Avian doppelgängers

The Willow Tit, greyish and small but with an unusually large head, also holds an unusual distinction: it was the last common British species to defy the scientists who were trying to identify every bird in Britain (although it has since become quite scarce). Until the late nineteenth century it was thought to be the same species as the Marsh Tit, which does look unusually similar, since it is small and greyish but with a slightly more normally proportioned head. Rather embarrassingly for the Brits, it was two German naturalists, Ernst Hartert and Otto Kleinschmidt, who recognised the mistake in 1897 after finding Willow Tits among the Marsh Tit specimens in the British Museum.

But these are not the only two bird species to look incredibly similar. The obvious question for all such birds is, 'If we can't tell them apart, how can they?'

In some cases, they can't – at least not all the time. Usually two species look very much alike because they are closely related, though in other cases the similarity has evolved for reasons that suit the bird. The Dusky Friarbird has a rather similar name to the Dusky-Brown Oriole, and this is no coincidence. They look like two peas in a pod – or like two Dusky Friarbirds (or, for that matter, like two Dusky-Brown Orioles). The oriole has evolved to look like the friarbird, whose forest habitat it shares in Indonesia, to reduce aggression by the latter.

But if we ponder how on earth birds can tell who is who (and therefore who is suitable mating material), we should reassure ourselves with the thought that many birds advertise themselves mainly by singing. In this they are a little different from us humans, though not completely different. Think of the hysterical female reaction to pop stars from Presley onwards, which echoed the response to the great nineteenth-century pianist and composer Franz Liszt, whose virtuoso skills on the keyboard – an extension of the human voice – made women swoon across Europe. Going back into prehistory, scientists have suggested that human language may have begun as a way to attract mates by showing off intelligence.

The cisticolas, whose stronghold is in Africa, are a case in point that shows how, for so many birds, voice is the primary means of attraction. There are about forty-five species of these small brown streaked birds – and since they are virtually all small, brown and streaked, it is very hard to tell them apart by sight alone. So you have to wait for them to sing – which is what cisticolas do too to work out who is who. Hence their wonderful names. There is Zitting Cisticola, named not after an acne problem but for the one-note sound it makes. There are also Wailing, Tinkling, Croaking and Rattling Cisticolas. There is even Wing-Snapping Cisticola, which tells other cisticolas which one it is through loudly clapping its wings together on the descending part of its display flight. There is also the Lazy Cisticola – a rather unfairly insulting name, since it actively cocks its tail a good deal (though I suppose this isn't as insulting as the Snoring Rail, which sounds as if it has taken lassitude to a new depth).

These names may seem rather over the top – much as 'bangers and mash' sounds more exotic if served up in a restaurant as *saucisses et purée de pommes de terre*. The Wailing Cisticola strikes you as very distinctive when you see its name on the page, but not in real life, where its song is very similar to the Grey-Backed's (though a bit more plaintive). Distinguishing them is so difficult that otherwise sunny dispositions can be affected in the attempt. A bird guide told me once of Kenya's Great Cisticola Rebellion, where his tour group decided that they weren't going to try to identify different cisticola species because it was too much of a headache.

But despite all these similarities, the birds that are most likely to try to mate with the wrong species do so not out of confusion over physical appearance, but for entirely different reasons. One is simple availability. Ducks in captive wildfowl collections often mate with each other, because they are thrown together by fate. A particularly common occurrence is the interbreeding of Tufted Ducks and pochards, even though they do not look very similar.

The near-impossibility of identifying some birds clashes painfully with the mindset of many birdwatchers, who should really perhaps be called bird identifiers, since that is the first thing they do before actually watching the birds. They find it frustrating that they cannot always do this, but should remember that this does not diminish the bird's beauty – it even adds an attractive air of mystery. Sometimes it's fun just to watch birds even if you don't know what they are.

So next time you're worrying about which bird is which, please console yourself with the thought that as long as *they* know, it's OK.

HOUSE SPARROW

Drab, dull but beloved

Britain's almost unique affection for the House Sparrow says much about our national character.

This small and rather dull brown bird is not one of nature's lookers. Many such birds, like the nightingale, compensate for their drabness with a beautiful song, but the House Sparrow merely contents itself with a cheeky chirruping. *Sparrers Can't Sing*, as the 1962 Joan Littlewood film about cockney life points out in its title.

In most of the world and through most of history, the sparrow has been held in a contempt that seemed to befit its dull and lowly status. The Ancient Egyptian House Sparrow hieroglyph represented things that were small, narrow or bad.

But London's East Enders always had an affection for the House Sparrow that lived right in the inner city of London. They gave it the ultimate accolade – almost unheard of for a bird – of its own name in cockney rhyming slang: 'bow' from 'bow and arrow' (because 'arrow' rhymes with 'sparrow'). The House Sparrow has been called into service for the opposite purpose too: as slang for something else. 'Barrow' can be called 'cock', from 'cock sparrow'.

When it was revealed in the 1990s that Britain's House Sparrow population was falling rapidly, it was not just East Enders who grieved. The news provoked an emotional outpouring much greater than that shown for many prettier, larger and generally more spectacular birds in a still worse plight. Why do we love the House Sparrow so much?

One reason is that it has for centuries been the most familiar bird of all to the average Briton. When we were toddlers and wandering around in the garden or at a park picnic, it was most likely a House Sparrow that our parents pointed out to us, when trying get our tiny infant brains round the concept of what a 'birdie' was. As the ubiquitous opportunistic feeder looking for scraps of food, it was the first bird we knew. This imbued us with an affection for it, but also a sense that, if the bird we knew best was doing OK, then so, surely, were we – and if it was struggling, would we be next? In worrying about House Sparrows, we are worrying about ourselves.

This is illustrated with great pathos in a story told to the US author John Steinbeck by a man who survived an air raid beside London's Hyde Park during the Blitz – a story included in Steinbeck's 1958 book *Once There Was a War*. A House Sparrow, hit by concussion, fell dead beside him. He picked it up, held it for a long time, and then took it home with him – thankful that he had survived, but sorrowful that the sparrow had not.

The second explanation for Britons' love affair with the sparrow is their innate affection for the ordinary, dingy and humble over the superficially impressive. The House Sparrow is the opposite of aloof, always popping up at our picnic tables to see what scraps are available. Its very dullness is appealing. The House Sparrow is the unadorned antithesis of the male bird of paradise, decked out in the most fantastical colours and performing the most convoluted display to impress the female. In Britain no one likes a show-off, so everyone likes a sparrow.

Contrast Britain's love of House Sparrows with Chairman Mao's Kill a Sparrow campaign in March 1958 – waged against the related Tree Sparrow, which also breeds in Britain despite perhaps the sharpest decline recently of any common breeding bird. The campaign was like many other birdbrained ideas that Mao had during his career – and doubtless everyone knew it was birdbrained at the time, but who can tell that to a dictator? Mao had decided that Tree Sparrows were agricultural pests, and declared a three-day war on them in all large cities. Three million people were enlisted to destroy the birds. The modus operandi was to gather in large crowds wherever there were sparrows, and then disturb them by banging gongs and drums. The surprised Tree Sparrows (and probably a good many surprised humans too) were prevented from feeding or even resting, and it was estimated that 800,000 birds were destroyed, though this must have included other hapless songbirds. The wheeze continued on a smaller scale throughout 1958, but was not repeated the next year, since grain yields plummeted and people starved. Sparrows eat crops such as oats and wheat, but also the insect pests that infest farmland.

In Britain, no one has fully got to the bottom of why House Sparrows are declining – though there are still about three million pairs left, so House Sparrows have simply made the change from superabundant to merely abundant. The oddest theory given is that mobile phones interfere with the bird's ability to navigate, and even to breed. But there is a consensus that changes in farming have played their part. The greater use of pesticides, for example, has killed many of the insects which sparrows need to feed their chicks.

Looking at the broad sweep of history, the House Sparrow is a bird whose golden era in Britain has gone – even though it is unlikely to disappear altogether. The arrival of agriculture thousands of years ago greatly expanded human numbers because it allowed for much greater population density than hunting and gathering; it allowed the House Sparrow to thrive too, as farming spread from Mesopotamia to cross the world, including the British Isles. In places such as North America, which the House Sparrow could not colonise naturally because this was too far to fly, it was introduced by settlers who wanted the familiar sparrow with them near their new homes. It has become possibly the most widespread bird in the world. In Britain, the peak of sparrow power was probably the era when farmland was pesticide-free and horses (which had to be fed oats, which sparrows love too) were used instead of cars. Go to any stable left today, and you will see House Sparrows hopping around looking for scraps.

It would be a pity, though, if House Sparrows were reduced, like stables themselves, to relics of a bygone age, holding out only in those places where people have the money to live in the past at weekends by mounting a horse and riding off into the fields. They survived the Blitz; let's hope they can survive the depredations of modern life.

WHITE'S THRUSH

The UFO bird

No bird utters a more eerie sound than the White's Thrush. On hearing its ghostly and plaintive one-note call, you will not immediately know that it is a bird. It sounds as if it could just as easily be a person sighing deep in the dark forests in which it is usually heard but rarely seen. Or, since sane people do not hang around in forests in the dead of night (though of course birdwatchers sometimes do), it could be a phantom bewailing its eternal mission to inhabit isolated and lonely places under cover of darkness.

But the White's Thrush sings for only the briefest of times, just before and just after sunrise. Unless you are a particularly early riser, by the time you get up it will have stopped, and you will wonder whether you have simply dreamt the sound rather than having heard it in those last few hours when sleep is more fitful and the calls of the night occasionally seep into your consciousness.

It is a pity that this bird that breeds in eastern Asia and Siberia is so rarely seen, since its appearance is rather spectacular. The bird's body has a beautiful mixture of orange and white crescents, edged in black.

The White's Thrush is associated with monsters in ancient Japanese folklore, but there is a modern postscript – its other-worldly voice has provoked UFO reports in Japan, or strictly speaking, UHO (Unidentified Heard Object) reports. Obviously some obsessive people who really need to get out more *are* getting out more and going to the forest, where they hear the thrush and report it as evidence of extraterrestrial visitation. But in a way this is quite understandable – the bird will often follow up a high single-note sound a few seconds later with a second, slightly lower sound, which has the air of a reply, as if two separate other-worldly beings are communicating with each other.

Because of the bird's association with dawn, it is a favourite of sound engineers working on Japanese samurai epics, who often insert its song into dawn scenes, when warriors are escaping under cover of the morning mist or waiting anxiously for the day's anticipated battle. It crops up in these dramas with a suspiciously impressive regularity that would be hard for the bird to match in real life. In much the same way, Tawny Owls are the staple of shady night-time scenes in British detective stories, even though in the real world their distribution is patchy and so their calls are only irregularly heard – and not at all in Ireland, the setting of a BBC radio play that provoked a complaint from a sharp-eared listener after an over-enthusiastic member of the production staff decided to inject a bit of owlish atmosphere into a night scene.

The voice is a key part of the White's Thrush's identity. Some ornithologists, many of them Japanese of course, argue that a race that lives on the Amami Islands in southern Japan should be considered a separate species because it has a completely different call – a rather unbecoming scratchy squawk. They have not yet won the victory in the fight to make the Amami race independent. Many of those birdwatchers whose raison d'être is to tick off as many species as they can are secretly relieved by this, since they are so hard to see. The Amami Islands are devilishly hard to get to, and even the more accessible birds live deep in the forest far away from the cafes of the islands' main town where you can drink cooling beers to recover from the oppressive humid heat. There are only about fifty of them, and they mainly call in the rainy season, when a trip to the islands could, if you are unlucky, be a complete washout.

The bird's striking appearance makes it particularly highly prized by British twitchers, even though on the rare occasions when it does make it to Britain, it is usually autumn and the bird has stopped singing.

However, the bird has a dog's dinner of a name. It is called after Gilbert White, the English country parson who, as we have heard, was the world's first birdwatcher, even though White had never seen the bird, which is only rarely found in England. Moreover, beginners to birdwatching usually confuse 'White's' with 'White', and expect a white bird. It just goes to show that you should never judge a bird by its name.

GALAPAGOS FINCHES

A little bird told him

Most of us know the Galapagos Islands as the location where Charles Darwin invented the theory of evolution, after looking at a group of species that had an extraordinarily wide range of bills but were otherwise similar. At least that's what we believe, but the reality of who invented what and when is a lot more complicated.

Darwin had his brainwave after studying a dozen or so species of finch on the Galapagos Islands, after travelling there on the HMS *Beagle* in the 1830s. Although all are about the same size, they range from the Large Ground-Finch, which can crack hard shells with its massive beak, to the Warbler Finch, which looks much more like a warbler than a finch because of a bill that appears as thin and sensitive as a pen nib, used to eat nectar, spiders and small insects. Collectively, they have become known by scientists as Darwin's Finches.

But what exactly was his brainwave? It was not, strictly speaking, the discovery of evolution. The notion that creatures would, generation by generation, slowly adapt more efficiently to their environments, had already been worked out by a few eminent thinkers in the generation before him, such as the Frenchman Jean-Baptiste Lamarck.

Darwin's claim to fame was more specific than that. It was his theory of natural selection – that within one species, the individuals that happened to be best equipped to deal with their environments were most likely to survive and reproduce, leaving individuals unlike them to die out. Over time, Darwin realised, these surviving individuals' descendants would become very different from other individuals that proved well adapted to a slightly different environment, and eventually they would be different enough to form a new species.

The impressively read among you will now ask that awkward question about Alfred Russel Wallace. Can this other English scientist not equally be said to have invented the theory of natural selection? Yes, but probably, to be fair to Darwin, slightly later. The evidence suggests that Darwin worked out natural selection in about 1838. But then, unwittingly observing John Lennon's dictum that 'Life is what happens to you while you're busy making other plans', got tied up with geological research (and possibly suffered from that common physiological affliction among many people who have discovered something radical – cold feet). So he didn't rush to publish his views until twenty years afterwards when Wallace wrote him a letter outlining the same idea (in the end, both men's views were presented to the world together, through a joint presentation in 1858). Darwin was clearly not a good man with deadlines – but Darwin's Finches eventually unleashed all manner of powerful forces, including the decline of religion in the West, the philosophical underpinnings of Nazism, and, in the view of many philosophers, a permanently more competitive spirit in virtually every aspect of human endeavour.

But let's not forget Patrick Matthew. If you haven't heard of him, this is because you are merely impressively, rather than impeccably, well read. Matthew was actually the first man to think of the idea of natural selection, expounding the theory in 1831 – the very year that the *Beagle* set out but before it reached the Galapagos chain. Matthew's mistake was to publish it in a book with the exceptionally dull name, *On Naval Timber and Arboriculture*. He had the genius to make the observation that since the Royal Navy cut down the best timber to make its warships, it was interfering with the natural selection by which the best trees would, over time, produce even better trees. However, not many people noticed this – and certainly not Darwin or Wallace.

What does this teach us about the confusing workings of humanity? On an intellectual level, it tells us that there comes a time when advances in scientific knowledge produce a small cluster of sufficiently intelligent people who all have the wherewithal to make the next advance. The invention of powered flight and the television show us that, and so too does natural selection. On a psychological level, it tells us that even men of the moral integrity of Darwin are reluctant to cede the gold medal on history's podium to another man. On a practical level, budding authors should remember never to allow publishers to put the word 'arboriculture' in the titles of their books.

And what of the finches? It turns out that they're not finches after all, but tanagers, a large family of birds in the Americas that look a bit like finches, with the same short, rounded wings. But Darwin's Tanagers does not have the same satisfyingly portentous dum-de dum-de rhythm as Darwin's Finches, so perhaps for that trivial reason the name has stuck. Even a genius like Darwin made little mistakes – which is quite reassuring for us ordinary people.

MAVERICK BIRDS

CUCKOO

The mafia bird

The evocative cuckoo (known, strictly speaking, as the Common Cuckoo) heralds the arrival of poetry as well as spring in the British Isles. Perhaps the first known poem written in Middle English – the forerunner to our modern language – is the anonymous thirteenth-century 'Cuckoo Song'. It begins simply and joyfully:

> Summer is a-coming in,
> Loud sings cuckoo!

The cuckoo is by no means the first of our spring birds to arrive, but it is the one with the most recognisable song of all. So far, it is the only bird sound I have managed to get my baby daughter to call back to me. The fact that you don't need to be a naturalist to recognise the cuckoo's call explains the truly nationwide competition, peculiar to Britain, to make the first report of it to *The Times* newspaper. This has led to a certain amount of chicanery that fits the bird's slightly shady reputation. A letter to *The Times* dated 12 February 1913 apologised for a letter to the same newspaper of 6 February 1913 that misreported the first cuckoo of spring, owing to the uncanny ability of a local bricklayer's labourer to mimic the bird.

Fewer birds have fascinated poets more than the cuckoo in the centuries since the 'Cuckoo Song'. Some have concentrated on its unwitting role as the harbinger of spring, but others have dwelt more on its naughty habit of laying its egg in another bird's nest, which allows the cuckoo to hightail it down to Africa for its winter hols much earlier than other migrants. John Milton's 1630s 'Sonnet to the Nightingale' is perhaps the best of all versifications of the cuckoo – exploiting with marvellous inventiveness the parallel myths that the nightingale signifies love, and the cuckoo a woman's capriciousness. According to folklore, every time the cuckoo sound was heard, a wife somewhere had just been unfaithful to her husband – hence the word 'cuckolded', used for a man whose wife has a tryst with another man. In reality, both birds arrive in Britain to serve as symbols of spring at roughly the same time in April. But in Milton's world the cuckoo always sings first. He implores the nightingale:

> Now timely sing, ere the rude bird of hate
> Foretell my hopeless doom, in some grove nigh;
> As thou from year to year hast sung too late
> For my relief, yet had'st no reason why.

One can only surmise that the twenty-something Milton had little success with women.

Cuckoos have fascinated scientists equally. How do they get away with such misbehaviour? First of all, they evolved to look like hawks, with a long thin tail and sharply pointed wings, to scare adult birds away from the nest so that they could lay their egg undisturbed. But why do the small birds which they exploit feed such a huge young bird that is so obviously not their own? One possibility is simply 'instinct'. Presented with a young bird begging for food, the small birds which the cuckoos use will, like automata, feed it.

Another possibility, known in learned circles as 'the mafia hypothesis', is that foster-parents are simply made an offer they can't refuse. This is based on experiments on species which practise the same technique as the cuckoo that breeds in Britain – 'brood parasitism', or 'dumping their eggs in other birds' nests'. It shows that at least two parasitical species, the Brown-Headed Cowbird of North America and the Great Spotted Cuckoo of southern Europe and western Asia, sometimes trash the nests of foster-parents that reject the parasite's eggs – making clear who's boss.

For all its wiliness, the cuckoo offers interesting proof of unfinished evolution. Female cuckoos specialise in taking advantage of different species. Cuckoos that use Reed Warblers will lay eggs that look just like the host family's. Some other cuckoos choose to lay eggs in a dunnock's nest – but in a strange anomaly, these cuckoos' eggs have not yet evolved to imitate the dunnock's bright blue specimens. Scientists assume this is because the cuckoo started exploiting the dunnock's hospitality more recently than other birds – though we know from a reference by Geoffrey Chaucer that the cuckoo was already taking advantage of the dunnock in the fourteenth century.

The cuckoo even made the reputation of Edward Jenner, the pioneer of vaccination. In 1789 he became a member of the Royal Society, the elite club for scientists, on the strength of his paper on the nesting habits of the cuckoo, a sort of eighteenth-century *CSI: Miami* in which he correctly documented to an incredulous public the fact that the baby cuckoo used its own body to eject the foster-parent's original and natural eggs from the nest. People had previously thought it was the adult female cuckoo that did this – and many continued to do so until film footage in the 1920s proved Jenner right.

However, let's not libel those saintly members of the cuckoo clan, such as the roadrunners, that do bring up their own young, and let's not forget that the fifty-seven species of cuckoo that impose themselves on other birds are not the only birds to practise brood parasitism. As well as the Brown-Headed Cowbirds I have already mentioned, Black-Headed Ducks do it too in the Americas, and some birds are even parasitical within their own species, such as the goldeneye duck, which sometimes lays its own eggs in another goldeneye's nest. Skulduggery is not confined to the cuckoo world.

MEGAPODE

The absent parent

You probably know a father who works feverishly hard at providing for the material needs of his children, but there is something missing from his relationship with them: he is perpetually absent.

You may be thinking of some alpha-male investment banker in Chelsea, but I'm thinking of the Australian Brush-Turkey.

The Australian Brush-Turkey is a member of a group of about twenty species of bird known as the megapodes, after the Greek for 'big foot'. It's strange that scientists should have the poverty of imagination not to name megapodes after their most interesting peculiarity: the fact that neither parent ever sees their offspring. I don't mean that metaphorically, but literally. Not referring to this in any way in this bird family's name is like meeting Brad Pitt and telling your friends afterwards that what you really noticed was that he speaks good French. That may well be the case, but it's not the most striking thing about him.

The Brush-Turkey, which looks very much like the American Turkey which we all know but is in no way related to it, adheres to this rather extreme form of child-rearing. The mother simply lays the eggs, but the father assigns himself, like all fathers who like tinkering with DIY, the task of building a huge mound of leaves and earth, sometimes 5 feet high and more than 10 feet wide, before the laying of the eggs. Occasionally he sticks his neck in it to make sure it's the right temperature – a heady heat of 33–35°C. But there is a serious purpose behind this outwardly eccentric behaviour: he is building an oven warm enough to incubate the eggs, without the body heat of mum or dad. By the time the newly born bird has hatched, the parents are far away, and they will never knowingly meet their progeny.

If a species is going to adopt a strange way of living, it cannot do it by halves, so this unusual system has forced Brush-Turkeys to develop a large set of other unusual characteristics. The most obvious necessity is that the young Brush-Turkeys must immediately be able to fend for themselves. This they can do, since they are able to feed immediately. Another is that the egg is large and contains a huge globule of yolk, which allows the chick to reach a fair size and strength. Young Brush-Turkeys are tough creatures. Some birds are altricial (blind, flightless and generally helpless in the days after they're born). Others are precocial (able to fend for themselves more quickly). But megapodes fit into their own unique category, which sounds like a line from a song in *Mary Poppins*: they are what scientists call 'superprecocial'.

Other types of megapode have habits that seem to most humans even stranger. Some bury their eggs in sand, and rely on solar power for incubation. Others take special tours to volcanic islands to put them in the hot soil there. These birds, buried deep below ground, sometimes struggle for days to reach the surface, unaided by their parents.

Birds are good parents in one specific and coldly unsentimental way: they are brilliant at producing enough young to ensure the survival of the species. But it seems, from our human viewpoint at least, a bond of duty rather than sentiment. The Golden Eagle, which patrols the skies above Scotland's mountains and moors, will lay two eggs. But the firstborn, given the head start by nature, will usually bully its sibling to ensure it secures the most food from the parents – even to the point of killing it. When the young sibling dies, as it usually does, from lack of food or fratricide, the Golden Eagle mother sometimes feeds it to the firstborn or even eats it herself. Once dead, its sole remaining function is as food. Nature is cruel, but efficient.

But there is a flaw in megapodes' strategy for survival: humans. There are more than twenty species of megapodes left, mainly in the Pacific, but it is thought that more than thirty may have gone extinct in the past few thousand years, of which the bulk seem to have died out before westerners even arrived – but after other humans did. This is perhaps the highest rate of extinction over the past few thousand years of any group of birds. Megapodes have been unusually affected by humans and the land mammals they have brought with them. As is so often the case in nature, the megapodes' advantages have been turned into disadvantages through the arrival of *Homo sapiens*. For example, the eggs are all the more attractive as food because they are so yolky, and the chicks are extremely vulnerable to introduced terrestrial predators in their first few days.

The same applies to seabirds such as penguins that nest in dense colonies: they have adapted to their forced close proximity by becoming extraordinarily tolerant of other penguins. However, this easygoing attitude towards creatures invading their personal space has also made them naively indifferent to human predation, with people testifying to penguins' lack of resistance when their eggs are plundered. It reminds me of the magnificent giant warrior in *Raiders of the Lost Ark*, who curls his sword above his head with a triumphant smile in preparation for carving up Indiana Jones – only to be despatched brusquely by the bemused hero's revolver. He is the perfect warrior for the age of swords, but swiftly rendered obsolete by a single gunshot.

But don't feel too sorry for the Australian Brush-Turkey, which, unusually for a megapode, is faring rather well. While many of its cousins have suffered because they are exploited by us and the animals we bring in our wake, this species has learnt how to be at the right end of exploitation. These formidable-looking birds commonly steal food from Antipodeans' picnic tables – often while the Antipodeans are still there. Think of it as The Revenge of the Megapodes (coming soon to a movie theatre near you).

HOATZIN

What walks like a bird and smells like a cow?

Meet the smelliest bird on earth.

The hoatzin of South America looks very much like a deeply surprised and rather suspicious chicken. It is usually found inside thick vegetation, its furtive eyes darting around beneath a floppy crest.

The hoatzin's is not a glamorous life. One local nickname is 'gypsy', but it seems a singularly inappropriate one. True, the hoatzin is multicoloured and has rather beautiful plumage, including a bright rufous wing-patch. But it is a weak flier, a poor swimmer, and even an ungainly climber. It certainly does not go a-wandering like a gypsy.

The final crippling blow to any delusions of glamour is its pong – described by people unlucky enough to get close enough as similar to that of a cow. Its alternative vernacular name, 'stinkbird', seems highly appropriate if a little rude. But scientists, inspired by its long crest, have compensated by giving it an extremely grandiose scientific name that makes it sound like a posh lady decked out for a night at the theatre – *Opisthocomus*, which means to wear one's hair behind one's head in a long manner.

The reference to a cow gives a clue to the smell's origin, and explains a lot else about the hoatzin's humdrum, sedentary existence. It subsists mainly on a diet of leaves. The bird eats a bunch of them, and then rests for a while to let the leaves settle. Since leaves are hard to digest, it has a huge foregut which exists for this purpose, much like the multicompartmental stomach of a cow. In both hoatzin and cow, the process of absorbing the vegetation is smelly.

The hoatzin's diet makes a mobile life both impossible and unnecessary. Impossible because, owing to the bird's heavy foregut and its contents, it is too heavy to fly well. Unnecessary because if you are eating leaves rather than trying to catch living prey, you don't need to move far or very fast.

The hoatzin's legendary smelliness has also turned out to be a happy accident of evolution. This rather large, pheasant-sized bird's inability to fly well could

have made it another dodo when European settlers arrived – easy meat and easily extirpated. However, its smell has prompted the European colonists who now make up most of South America's population to assume that it will not add up to an appetising meal, despite the fact that some indigenous tribes eat it quite happily, and others consume the eggs, which aren't smelly at all. The open-minded should note that beef tastes delicious.

In its bid to be spared by us the bird has another ace up its sleeve, or whatever stinking orifice serves as a substitute. It has an ability to eat leaves that are toxic to other animals, with the aid of friendly microbes in its body. If fully understood by scientists, this advantage could be transferred to cows – allowing them to eat a wider variety of foliage than they can at present. A larger global population of cattle would help stave off world hunger (though we would have to reduce the corresponding increase in the methane greenhouse gas produced by belching, flatulent cows – a problem on which New Zealand's scientists are gamely working, pegs on noses). Here is the biodiversity argument in action: that it is in our own self-interest to preserve as many species as possible because they could be helpful to us, and few more so than this bird. The exploitation of birds that have uses to us can help save them as long as we do not over-exploit them at an unsustainable rate. For example, the exceptionally warm, waterproof feathers of the Eider Duck can be used for eiderdowns, and pheasants can even be bred for hunting in the British countryside, as long as we avoid the common human tendency towards unsustainable excess, which is just as apparent in our tendency to cause banking booms and busts as it is in our relationship with birds.

It would be a pity, too, if the hoatzin became extinct before we found out exactly where it fits in the bird kingdom. Since its discovery by scientists in the year the United States declared independence, scientists have placed it in or near at least ten different families, including the rails, the pigeons and the cuckoos. Recent DNA analysis that attempted to resolve the issue once and for all has only deepened the debate, showing that scientific advances often exacerbate controversies rather than resolve them. The current fashion is to say that the hoatzin should be in its own scientific family, because it is not very closely related to anything else on this earth.

The reader may think I have been unduly harsh on the hoatzin, whose social habits do after all have the sweet smell of success about them. Observers brave enough to study the bird at close quarters – presumably taking care to stay upwind – notice the bird possesses the twin virtues of monogamy and helpfulness. Hoatzins form little colonies of up to eight birds that all bear their share of looking after the young by defending territory, building nests, feeding fledglings and even incubating them when necessary. The Guyanese are certainly happy to be associated with it – it is the country's national bird. The hoatzin may lack the majesty of the Bald Eagle, its US equivalent, but perhaps its helpfulness and homeliness make it the more likeable bird for all its defects.

CORNCRAKE

The bird that plays hard to get

Once in a while history throws up a great figure who seems to come from nowhere. Nothing in their background indicates future brilliance. There are two common reactions: conspiracy theorists attribute their achievements to someone else; the bien-pensants choose to ignore them. William Shakespeare, the son of a glover in rural Warwickshire, has frequently suffered the former fate. John Clare (1793–1864), a rural labourer from rustic Northamptonshire who was both a first-rate poet and a pioneer in the world of ornithology, endured the latter.

After limited initial success, Clare was largely ignored as a poet until the early twentieth century, well after his death. Now, however, he is regarded by many as among the greatest English poets of his time. Many of his works were inspired by nature, and he probably wrote more about birds than any other poet of his calibre has in history. His pieces about birds are almost like rhyming field guides, perfectly encapsulating the key features of each species, and 'The Landrail' – an old name for the corncrake – does the job best of all.

The most striking aspect of the fifteen-verse poem is that it does not once mention the bird's plumage. You may think this a dire dereliction of duty, for a poet trying to describe a bird. But Clare has identified the key point about the corncrake: however much it calls (and it loves calling, with a two-note sound likened to drawing a comb across a matchbox), you will almost certainly never see it. The poem reflects this perfectly: the corncrake is teasing the poet, because it never shows itself, and the poet is teasing us, because he never shows us the corncrake. Instead Clare spends most of the poem describing the experience of looking for this 'living doubt', as he memorably describes the elusive bird:

> They look in every tuft of grass
> That's in their rambles met
> They peep in every bush they pass
> And none the wiser get.

We should not be too disappointed if we, like Clare, fail to see the corncrake. Like the slightly less secretive nightingale, it is a small brown bird of unspectacular appearance, though of a different shape. The corncrake is a kind of rail – like the Water Rail or moorhen – but is a maverick among rails since, as the old name implies, it eschews water for farmland. It is also, like cops in New York crime dramas, a maverick with a complicated personal life – males often mate with two or more females, which nest near to each other.

In the end Clare went mad because he was unable to cope with the huge changes wrought on his beloved Northamptonshire landscape during his lifetime, including the ploughing up of pastureland and the drainage of the fens he loved. He was much like a bird that is sensitive to its ecosystem and unable to cope with changes to it. Many of the birds he recorded in Northamptonshire, including the corncrake and the reed-loving bittern, have long since disappeared as regular breeders in the area because of changes to habitat or farming practices. When machines rather than men with scythes started to cut fields in the nineteenth century, the corncrake suffered.

Young birds could not run fast enough to escape. Within the British Isles, the corncrake is now mainly a bird of carefully managed nature reserves in Scotland and Ireland.

Clare's bird sightings – some of them recorded through his poems – are extremely valuable historically. He was one of the first men in history to make an attempt to record his local birds as a modern birdwatcher would today – to observe them about their daily business in acute detail and then to pass on, leaving them unharmed. Those of his contemporaries who were interested in birds – and there were not many for much of his life – were more prone to shooting them.

Are there men like John Clare around today? Yes there are, in the shape of the tens of thousands of birdwatchers who have what in birding jargon they call a 'local patch' – an area near home which they visit at least every week, and sometimes every day, in an effort to retain their sense of belonging to the particular stretch of land they inhabit. Poets are no different from the rest of us: they share the same feelings, but deviate only in their genius for putting them to paper.

The satisfaction of the local patch is probably best described by the Victorian poet Wilfrid Scawen Blunt. The Old Squire in the eponymous poem spells out movingly the joys of all that is regular but not mundane, all that is seasonal but in essence unvarying, in a poem full of bird references:

> I covet not a wider range
> Than these dear manors give;
> I take my pleasures without change,
> And as I lived I live.

There is, though, a twist in the tail. Of the five wild birds he mentions to put his case for a never-changing England, one is the pheasant, which is not a native bird at all, but was introduced by the Normans – a reminder that even what we love as old is often newer than we think.

As a footnote, if you think spying on a corncrake is difficult, try staking out a quail. It is almost never seen – to the point where one wonders how naturalists actually know what a quail looks like. In its eternal effort to avoid human contact, it has an added weapon – what scientists call ventriloquism. The quail can throw its voice so that it sounds as if it is coming from somewhere else. Keith Harris did not need to pass through all the palaver of learning how to transfer his sounds to Orville the Duck. He could just have used a quail, which would have done all the work for him.

NIGHTJAR

The snatcher of babies

Few birds have been mistrusted, feared and positively reviled in history as much as the nightjar.

The hatred of the bird has a long history. Aristotle recorded that the European Nightjar – the species found in Britain – made livestock go blind by sucking their milk, which explains the Latin name *Caprimulgus* given to one branch of the family, meaning 'goat sucker'. Although the name is an inaccurate slur that should prompt writs from the nightjars' lawyers, it has stuck – and so has the nightjar's evil reputation.

In many parts of North America the Whip-Poor-Will, a local species of nightjar which owes its bizarre name to its call, has been associated with death. In Sulawesi it was thought that the call made by Heinrich's Nightjar, known locally as the Satanic Nightjar, was actually the noise of it sucking out people's eyes (though this begs the question, how on earth did they know what that sounded like?). In parts of England the nightjar is associated with Puck, the usually malign nature spirit who appears in *A Midsummer Night's Dream* (though Shakespeare subverts the legend by breaking with tradition and making him rather likeable). Some Aborigines in Australia thought the

Spotted Nightjar snatched babies away under cover of darkness. In short, people across an enormous range of cultures seem to have reacted to the nightjar in an uncompromisingly negative manner.

As it muses on its less-than-perfect press, the nightjar must be thinking to itself, 'Was it something I said?' Its very voice has cemented the nightjar's reputation as an other-wordly bird. Some of the *Caprimulgi* nightjars make a sound like a death rattle with their mouths, and the males produce a noise like a whip cracking with their wings when trying to attract a mate (though if she believed even half the legends, would any sane female nightjar think of settling down with a male and bearing children?). Their sometimes huge eyes on generally very round heads give them an oddly human appearance. This, and the habit of some nightjar species of sating their curiosity by flying a full circuit round nearby humans as if they are trying to communicate with

us, is perhaps the origin of a legend that their bodies contained the souls of unbaptised children. The bad associations of nightjars have, of course, developed their own momentum. Thus the onomatopoeic call of one North American species gives us the name Chuck-Will's-Widow – even though the rhythm could just as easily be rendered as something cheery like hi-hi-hello.

It should come as no shock, then, that nightjars have seeped into popular culture's less wholesome nooks. The horror writer H. P. Lovecraft uses the Whip-Poor-Will as a plot device in his 1929 story *The Dunwich Horror*. With such opprobrium heaped upon it by Lovecraft and legends, it is small wonder that the country and western singer Hank Williams Sr described the nightjar in his song 'I'm So Lonesome I Could Cry' as sounding 'too blue to fly'. It would make a good newspaper headline: 'Famous country singer claims bird suffers clinical depression.'

The nightjar, which feeds on night-flying insects, reflects humanity's complex attitude towards the dark – a time when, in societies that pre-dated the electric light bulb, few people doing anything respectable would be up and about.

But the nightjar adds an extra element of mystery that other night birds such as owls do not. These perfectly camouflaged birds, whose grey-brown and white plumage makes them merge perfectly with branches and dead leaves during their daytime roosts, are poorly known – and for birds, as for other humans, mystery creates mistrust. Nightjars are the avian equivalent of the stranger in town in a western. We have the habit of projecting our own fears, hopes and suspicions onto unknown birds just as much as onto strangers. Our attitude towards nightjars is simply the flip side of thinking we have fallen in love with a dark good-looking stranger on the tube when we are not old enough to know better (and indeed sometimes even when we are old enough to know better).

Nightjars continue to be so mysterious that I would wager there is at least one species that we haven't yet found, in addition to the eighty-five or so that we do know about. Of those eighty-five, we know of some only through pure chance, and their feeding, breeding and other habits still lie outside our knowledge. Until 1996 only one Heinrich's Nightjar had ever been seen by scientists, though as is often the case it appears to have been known by the locals. Vaurie's Nightjar is still waiting to be relocated after the initial discovery of a single bird in 1929. It is found, if that is the right word, in a place as near to the ends of the earth as you can get without going to the polar regions – the Taklimakan Desert of south-western Xinjiang in China.

Can anything be done for the nightjar's reputation? Time for its lawyers to call in a PR guru. Max Clifford perhaps?

BEE-EATER

Twitching – hobby or addiction?

Bee-eaters are often the birds that people most desperately want to see after eagerly rifling through those lists, sent out in advance by purveyors of foreign birdwatching holidays to whet appetites, of what might be encountered on an ornithological odyssey.

A look in any field guide explains why. The different bee-eater species are invariably decked out in a range of bright colours, with two splendid flourishes at either end to complete their beauty: elegant long tail streamers at the back, and a long, sinuous bill at the front. Every birdwatcher who has seen bee-eaters has his or her favourite. Mine is the Southern Carmine of southern Africa, which casts aside any pretence of understated beauty by simply looking heart-stoppingly gorgeous. At one end of the subtlety spectrum is the fair Bachman's Sparrow of the United States – one of those small brown birds that looks rather boring at first, unless you can find the time to examine the subtle streaking on its neck, back and wings. At the other end of the spectrum is the Southern Carmine Bee-eater. Its overall pink colouring, long tail streamers and long curved bill are immediately arresting, though the turquoise, blue and green finishing touches to its plumage are equally beautiful.

The behaviour of bee-eaters is as compelling as their appearance. True to their name, they eat bees. (This might sound a trite truism, but many birds have monikers that are downright misleading. The oystercatcher very rarely catches oysters, the Honey Buzzard isn't a buzzard, and the Woodland Kingfisher prefers insects to fish, which is probably a good thing since it also chooses woods over water. For more on just how odd birds' names can be, see the Arctic Tern essay.) Bee-eaters have a rather brutal habit of removing the bees' poisonous sting by using their long bills to bash the bee repeatedly against a hard surface such as a twig. It is fascinating to watch them catch a bee and then painstakingly go through the process of preparing their dinner, with the diligence of a cordon bleu chef.

It was birdwatchers' mania for seeing bee-eaters that helped turn the 1955 search by Britons for a pair of European Bee-eaters nesting near Brighton into what some see as the world's first twitch – an epochal event in birdwatching history.

First I should explain what a twitch is, and just as importantly, what it is not. It means travelling to see a bird which is only rarely found in your country, but has turned up there by chance. In the European Bee-eater's case, this is because they are prone to what experts call 'overshooting' – migrating from Africa with the aim of breeding in mainland Europe, but ending up a little too far north, in Britain. Any hiker can wryly testify to how easy it is to overshoot, though it would show an unusual degree of incompetence to hike all the way into the wrong country. Twitching does not mean going birdwatching in general, although the media sometimes assumes this. This is an example of the common phenomenon of people trying to show they know more than they do about a particular social group by using its lingo, but getting the lingo wrong.

This proto-twitch for bee-eaters unfolded at the leisured pace of 1950s society, in contrast with today's fast-paced, high-tech twitching. When a local birdwatcher found the birds, he informed all his contacts by post, and many took weeks to see them. Twitching has moved on since then. In the 1970s and

1980s, when telephones had become ubiquitous, birdwatchers used to phone their friends for news of twitchable birds. By the time I spent two weeks as a teenage twitcher in the early 1990s on the Isles of Scilly (a hotspot for American birds of extreme rarity), a few twitchers were using walkie-talkies to chat to friends a hundred yards away about the bird's precise location. Mobile phones eventually superseded walkie-talkies, and nowadays you can get up-to-the-minute news of where rare birds are (and just as importantly, whether they are still there and the long trip is worth making) on your BlackBerry, through special subscription services. The bee-eater twitch was a long way from this, but the hallmark of twitching was there – going the necessary distance, however far it is, to look at a bird that shouldn't be there and was discovered by someone you hardly or do not even remotely know.

Why did twitching become popular in the first place? It fulfilled one of the motivations behind ordinary, everyday birdwatching – the hunting instinct of finding one's target bird, raising one's binoculars to it much as one would the sights of a gun, and then, with that sense of 'got it!', securing the necessary look at the diagnostic features that mark it out as one species rather than another. In hunting, the most frustrating thing is to find one's quarry and then to miss – to be so close to success but then to throw it away. The twitcher's worst nightmare is exactly the same, though it has its own jargon. It is the UTV or 'untickable view' – to see the bird but not well enough to tick it off as that particular species. It is, therefore, really a resurfacing of the old Victorian collectors' instinct for shooting and collecting rare birds – suggesting that humans' attitudes towards birds have not, after all, changed as much as is sometimes supposed.

Even victory can be brief – after hundreds of miles' worth of guzzled petrol and hours stooped over a telescope until your neck feels like a block of wood, the only view you may get all day is a five-minute glimpse, before the bird disappears once more into the undergrowth. As Lord Chesterfield said – allegedly about sex, though I think he was talking about twitching – 'the pleasure is momentary, the position ridiculous, and the expense damnable.'

If that doesn't sound to you like a fun day out, remember that it is only a minority pursuit even among birdwatchers. There are plenty of more civilised ways to enjoy birds (though you can expect blank looks from any twitchers whom you say this to).

GREAT SPOTTED WOODPECKER

The bird with the built-in shock absorber

Britons commonly observe that everything was better in the old days – whether it be people's manners, the pace of life, or the cleanliness of streets. Birds are no exception to this old saying.

But it is a little-known fact – even among birdwatchers – that, since the mid 1990s, there have been more winners than losers. Surveys of common breeding birds since 1995 show that fifty-eight have increased, against forty-four that have fallen, and for five it is difficult to tell one way or the other. The Great Spotted Woodpecker is a perfect example of a species that has thrived. Numbers have approximately doubled during the period to about 40,000 pairs. All nature lovers adore woodpeckers because of their unusual appearance and fascinating habit of noisily drumming their bills on trees to peck holes for nests and to attract mates, aided by shock-absorbent tissue in the head that has been studied by the makers of motorbike helmets. But that special thrill I had of seeing one in my childhood has rather paled because the Great Spotted – already at that time the most common of our breeding woodpeckers – is becoming even commoner. On a day's walk I can usually expect to see one, if I listen out for its one- or two-note call that sounds a touch like a champagne cork escaping from a bottle. Seeing a woodpecker is nice, but certainly no longer calls for a bottle of bubbly these days. If it did, nature lovers would be pretty drunk by now. There are birds that have thrived more even than the Great Spotted – numbers of the stonechat, a small songbird that often obligingly perches on the very top of bushes as if to say, 'Hey, look how well I'm prospering, you can see me everywhere these days', have soared by 168 per cent.

The trend for British birds to increase rather than decrease is even more marked when it comes to rare birds. There are few rare breeding birds which conservationists have failed to rescue from extinction in Britain over the past two decades. Thriving examples like the bittern and avocet are much more

common than exceptions like the highly camouflaged wryneck, which is actually another kind of woodpecker although it looks more like a nightjar – or, to be honest, most of all like a lump of wood. The wryneck now breeds only fitfully in Britain.

Why is this? For rare birds, the answer is relatively easy. We simply know a lot more than we did – both about conservation in general, and about what is good for particular birds. So we always understood that the bittern, for example, liked large reed beds. After some decades of creating large reed beds without great success, we worked out that they also like a patchwork of water channels, clear of reeds, to patrol.

For common birds, the answer is a little harder. One argument is that we have a lot more reserves than before. They can never cover more than a minority of the countryside, but are enough to stop bird numbers from falling even further below their levels in Victorian times – and indeed to start them rising again slightly. Another argument is that gardens are also nature reserves in all but name, since they provide good, well-protected habitat, and when put together each small garden accounts for a huge area. In particular, Britons' winter feeding of garden birds prevents numbers from falling too far in the winter.

But is there a particular reason why the Great Spotted Woodpecker is thriving? A common theory is that the Dutch elm disease that struck so hard in the 1960s created just the sort of dead trees which the species loves. In nature, what is bad for one living thing is often good for another.

HAIRY WOODPECKER

Haven't we met before, Mr Audubon?

The Hairy Woodpecker looks much like our Great Spotted, with a black-and-white criss-cross pattern across its body. In the field, this bird of North and Central America looks even more like the Downy Woodpecker. The Hairy and the Downy sound almost opposite by name, leading the novice birdwatcher to think they must be the easiest birds of all to identify in the woods. However, their striking names are based on the examination of the skins of specimens in museums, rather than in the field (where they have to be studied closely to tell them apart). In the wild the Hairy does not look hairier than the Downy, and is best distinguished from it by its longer bill.

The two birds are not particularly closely related, but are good examples of convergent evolution, where two birds grow over time to resemble each other closely in response to similar needs – much like swallows and swifts (even though swallows' closest relatives are probably the tits, and swifts are cousins of hummingbirds). The Hairy Woodpecker is a common bird in deciduous forests, but is perhaps most common of all in the writings of John James Audubon, the idiosyncratic founder of American ornithology, who during his roller coaster of a career named it on five different occasions after five different people. Audubon also fell into the trap on more than one occasion of 'discovering' new birds that had already been discovered by someone else under a different name. Why is a person with such a disorganised approach to finding and naming species seen as the father of American ornithology?

Audubon was in many respects a deeply incompetent man. His career began inauspiciously when he tried to follow in his father's footsteps by embarking on a life on the ocean wave, only to drop out on finding himself prone to seasickness. Then on going into business he proved so spectacularly unsuccessful that he managed to go bankrupt – a mishap that landed him in jail. In fact, he failed to achieve any worldly success until his forties – by which age many men of his time (1785–1851) would already have died. He fell into depression after rats ate his bird drawings, and caught yellow fever and nearly died after falling into a creek while out in the countryside, but his long-suffering and ever-loyal wife Lucy (whose account of life with her hapless husband I would love to have heard) nursed him back to health.

But his drawings of birds are the most impressive I have ever seen. Each of the paintings in his magnum opus, *Birds of America*, which came out between 1827 and 1838, are 39 x 26 inches – a new folio christened 'elephant size', designed to fulfil his dream of drawing each bird to its actual dimensions. It is the combination of the size and beauty of the drawings that impresses. Imagine standing a foot away from a Tricoloured Heron that is even more striking than the real thing at that range. This impressive tome has become the world's most expensive printed book, with the most recently sold copy changing hands for £7.3m.

To be fair to Audubon, his career until *Birds of America* – first published when he was forty-two – reads less like the chronicle of a complete dolt, and more like that of a man distracted from daily life by his single-minded passion for one thing: birds.

He developed an early love for birds while growing up in Haiti on his father's sugar plantation, and never really got them out of his head. Audubon took any opportunity he could to spend time looking at them (and, given that people did not have the binoculars to study their fine plumage details at the time, often shooting specimens). He neglected business on his father's plantation to sketch birds.

Audubon was an eccentric with a vision: to chronicle the birds of the United States more completely, and more beautifully, than had ever been done before. He was also rather an obsessive perfectionist – constantly destroying his drawings to force himself to draw better ones. Audubon's drawings are considered very mannered and rather unrealistic by modern sketchers of birds, despite his use of wires in an attempt to put them in naturalistic positions, ready for sketching. But whatever people say about the pictures now, they are works of art, and they gave American ornithology an elephantine kick-start. Sometimes the world needs eccentrics like Audubon – inspired eccentrics are akin to the aberrant members of a bird species that drive it forward and continue the process of evolution.

Audubon's book is also a memorial to a vanished age, since no fewer than six of the birds that he drew have since been declared definitely or possibly extinct. The Great Auk is definitely gone, but the Eskimo Curlew might just be hanging on by a thread, somewhere within the huge and isolated swathes of Arctic Canada and Alaska where it used to breed by the million. Intriguingly, there have been reports of the bird from reputable birdwatchers as recently as the 1980s. You can check out a life-sized version of the Eskimo Curlew, by a strange quirk of fate, in the Isles of Scilly Museum – a specimen shot on the island in 1887. The Ivory-Billed Woodpecker is also probably gone, despite recent unsubstantiated claims that it has been seen in the wilds of Arkansas.

But what of the Hairy Woodpecker? It is thriving, with more than nine million birds across North and Central America, thanks in no little part to the Audubon Societies set up across the United States in honour of John James Audubon. Let's praise one of the most gifted incompetents in the history of humanity.

NORTHERN BALD IBIS

The bird that existed after all

The Northern Bald Ibis shows us that history matters – and that birds are no exception to this rule.

Conrad Gesner, a Swiss doctor who published four encyclopedic volumes on *The Histories of the Animals* in the sixteenth century, was not the most reliable of witnesses. His book includes animals and birds we know about like the rhinoceros, but also rather more fantastical beasts such as the unicorn. So when, a couple of centuries later, people read Gesner's description of a bird 'larger than a hen, with black plumage, a naked face and a long bill' that lived in the mountains of Switzerland, but could see no bird like it, many decided to write it off as another of Gesner's flights of fancy. It sounded just the sort of make-believe creature that someone with an excessively vivid imagination would create – a bird with a bare face like a human's. Matters were not helped by the fact that no one could decide what kind of bird it was. Gesner thought – incorrectly – that it was a sort of raven. The eighteenth-century Swedish naturalist Linnaeus believed – even more incorrectly, if that's possible – that it was related to the hoopoe, a songbird that looks a bit like a jay. To be fair to Linnaeus, in Gesner's crude woodcut of the bird, it does rather resemble a hoopoe.

But then, in the early twentieth century, birds fitting this description turned up in North Africa and the Middle East. Fossils were also found in Europe, showing that Gesner had been right all along. The bird was simply forgotten about on the continent after becoming extinct there about a century after Gesner recorded it for an incredulous posterity, probably wiped out by hunters for its delicious meat.

It matters now that we know it lived in Switzerland and other parts of central Europe – just as Gesner had said – because the Northern Bald Ibis is endangered again. Several attempts to breed the bird in captivity before reintroducing it into the wild are now under way, and some of these trials are taking place in Europe, based on the assumption that if the birds knew that Europe was suitable for them hundreds of years ago, who are we to argue with them? The history of

birds can tell us more about suitable breeding grounds than any theoretical guesses by scientists, because birds' preferences are so complex that if we simply guess where might be good for them, we are likely to get it wrong.

One European country that has two reintroduction projects is Austria, where further historical research shows that it became one the world's earliest birds to enjoy official protection, on the orders of Archbishop Leonard of Salzburg in 1504 – though one hopes people listened to his sermons more than they did to his decrees on conservation, which failed to prevent the bird's European extinction. The knowledge that Austria is a former homeland to the Northern Bald Ibis has given the country ecological as well as moral grounds for trying to re-establish it there.

Once you start looking for a bird in the annals of history, references to it crop up much more regularly than you might have imagined. We now understand from a fifteenth-century Spanish falconry book that the Northern Bald Ibis used to breed in that country too. Reintroductions have been made in Spain with the helping hand of the military, which like most armed forces has large areas of sparsely populated land for training. The Great Bustard's abortive 1970s British reintroduction likewise used British army property on Salisbury Plain.

The Northern Bald Ibis's revival has probably been aided by a spectacular ugliness that almost makes it beautiful. This has made it a favourite with zoos, which have more than 1,000 of the birds – far more than there are in the wild. As well as zoos, which have served as sources for reintroductions into natural habitat, the full force of ultra-modern technology has been marshalled to their aid. These include the satellite tracking of the surveillance state, which usefully if rather spookily established that some of the ibises had settled down for the winter in suitable habitat in Ethiopia – where they were seen by human eyes only later. This momentous example shows how bird recording has moved in the twenty-first century not merely beyond shooting, the method up to the end of the nineteenth century, but also beyond sight records of live birds, the method of the twentieth.

Gesner's verbal description of the bird turned out to be pretty accurate – much more accurate in fact than his picture. But it remains an enigmatic creature. Most ibises like water habitats, but the Northern Bald Ibis breeds on cliffs and has a propensity for feeding on dry steppe land. Looking at the case of the Northern Bald Ibis, I wonder how many real-life birds of the past, dismissed as inaccurate fantasies when described in dusty old books, have gone extinct altogether.

BEE HUMMINGBIRD

Loved to death

The hummingbird is the bird that, when spied for the first time in your life, looks least like one. Before you train your binoculars on this hovering object, it resembles a type of insect – a bee perhaps.

Some hummingbirds are not much bigger than bees. The appositely named Bee Hummingbird of Cuba is the smallest bird in the world, at only 5 or 6 centimetres long. This is only a little more than some of the larger bumblebees, which can approach 4 centimetres. The similarity is more than superficial, since in common with bees, they feed on nectar – the highly concentrated sugar provided by plants.

Hummingbird fans might worry that these tiny birds will be tired out by all that flapping, like little children who cannot pace themselves. For hummingbirds, however, it makes sense. They have to hover – at up to eighty wingbeats a second for some species – to remain motionless while performing the tricky mechanical feat of sucking the nectar out of plants with their long tongues, sheathed in long bills. (It is no wonder that Igor Sikorksy, the Russian-American who pioneered the helicopter, said he thought up many of his ideas for it from watching hummingbirds.) This process allows them to absorb enormous amounts of energy, but it also requires a lot of oomph – because hummingbirds have to visit hundreds of flowers a day.

This rather extreme lifestyle produces some odd accidental by-products. One is the bee-like humming sound made by their wings – hence their name. Another is their constant urination, as they try to get rid of all the water they are absorbing from the nectar. You don't want to get too close to a hummingbird, even though they are so pretty.

The burning question about hummingbirds is, why exactly are they so beautiful? The most common colour in the bird world is almost an anti-colour – the brown hue of camouflage, of inconspicuousness, of minding your own business and not attracting attention.

The modern explanation for hummingbird pulchritude is that they can afford to be beautiful because they are small. They are the bird world's equivalent of cucumbers – predators would use up more energy in taking and eating these tiny acrobats than they would absorb from them. So the natural evolutionary tendency for males to show off as much as they can get away with (and we're not just talking about birds) is given full rein for hummingbirds – they can become as multicoloured and conspicuous as they like, in order to attract females.

An alternative explanation is supplied by a charming Mayan legend, which is also rather logical for a society that did not know about evolution. The Mayans said the head of the gods had an assortment of tiny bits and bobs of material left over, after making all the birds, and that he used them to create the multicoloured hummingbirds.

Humans have been fascinated with these creatures for as long as they have known them. The Aztecs even had a hummingbird God – which encapsulates the sense of awe with which they have been held. 'I saw a hummingbird and I was its captive', a friend wrote to me once on seeing her first.

Captivity has worked both ways. The Aztecs bred hummingbirds in enclosures for hundreds of years, so they could use their feathers for sundry religious ceremonies. While the Vikings thought that warriors who died bravely in battle went to Valhalla – Viking heaven – the Aztecs believed that they were reincarnated as hummingbirds, and spent eternity sipping nectar.

But the taking of these New World creatures on an industrial, unsustainable scale arrived with the Europeans, who were entranced by these birds and gave them exotic, resplendent names: no mere 'Greater Spotteds' and 'Lesser Spotteds', but titles that sounded less like bird names and more like cocktails. Jamaican Mango for anyone? These Europeans used hummingbird feathers in their millions to decorate ladies' hats and clothes, although nowadays habitat loss is a greater peril than hunting for endangered species such as the beautiful Oaxaca Hummingbird of Mexico.

In Victorian times hummingbirds were like small collectable antiques, so they were now killed in their millions for more than just their feathers. John Gould, the highly entrepreneurial and intensely ambitious nineteenth-century British naturalist who became a rich man through his various bird-related activities, was obsessed by them. Although he did not see his first live hummingbirds until he went to the New World at the age of fifty-three, he admitted to having dreamt for years about them zipping around in their flower-rich habitat. The ultimate incarnation of his obsession was the mammoth undertaking of displaying 1,500 specimens, mounted on wires to render the verisimilitude of hovering over flowers, to the paying public at the London Zoological Gardens in 1851 (though Gould did not actually see a live hummingbird until six years later). To the modern nature lover it sounds macabre and deeply tasteless, but not to Victorians, who paid to see this in droves. They loved hummingbirds to death.

SWIFT

The bird with no feet

Is the Common Swift, as it is more correctly called, the most birdlike of all the world's 10,000 or so birds?

It seems a bold claim to make. But if birds are noted and treasured most for their ability to fly, this pointed creature that resembles a flying scimitar takes it to extremes. The swift, which breeds in Britain and most of the rest of northern Europe, is the only species that we know for sure sometimes sleeps while flying – it has even bumped into planes while doing so. The swift combines sleep with flight through a mixture of gliding and slow wingbeats, usually at between 1,000 and 2,000 metres high. We do not know all the details of how it manages this, but we do know that swifts do not sleep with the same intense depth as humans – suggesting that even while slumbering, they have some consciousness of the world. Dolphins have a similar ability, and can swim slowly while sleeping.

The swift's Latin scientific name, *Apus apus* – 'no feet no feet' – confirms a feature that emphasises its flying ability: it has feet, but only just. The swift is built for flying – and can even have sex in the air, which is more than most of us can manage. However, its extremely short feet are good for clinging on to nesting sites on cliffs and the walls of houses, though they are not able to do much more. The swift's lack of feet has ensured its immortality in the British coats of arms originally created for some younger sons in noble families, symbolising the fact that, according to custom, little brothers were not entitled to a footing in the family lands and had to make their own way in the world.

Birds' power of flight appeals to humans largely because of the freedom from earthly constraints that it represents. This liberty was painfully apparent to me when, as a teenager, I used to listen simultaneously to the headmaster droning on during the morning assemblies that started the day in the Gothic Great Hall of my school – built with the sturdiness of a prison – and the delighted shrieking of the swifts circling somewhere above our heads outside. Their reputation as the late risers of the bird world also appealed to me at that sleepy age. Because swifts prey on insects, there is not much point in getting up before the insects do, and the insects often need a few hours after daybreak for the warmth of the sun to make them active. For this reason, swifts are particularly prone to listless inactivity on wet mornings – much like humans when they have the chance.

There is another swift predicament with which any teenager who has ever been told off for having an unkempt room can empathise. Swifts suffer from the modern world's increasing tendency to want everything tidied up. Over the centuries they have taken with relish to nesting on the walls of buildings – which are, after all, much like the cliffs that they used before. So although they seem very other-worldly to us, they are unusually dependent on us at that crucial time in a bird's life when it feels the time has arrived to start a family. But busy builders plug gaps in eaves, put concrete fillets under tiles, and generally contribute to the unnecessarily extreme neatness and conformity of the world. They are the enemies of swifts, whose numbers have declined in Britain. In this respect swifts are like European vultures, which have suffered from rules and regulations encouraging farmers to incinerate their dead livestock rather than leave their corpses to be pecked away by these airborne dispatchers of carrion.

But help is at hand in the form of swift nest boxes, which swift lovers can buy and attach to walls. For budding conservationists who do not have gardens, they are one of the few ways in which they can use their rather limited domains to help birds. Swift nest boxes have gained a popularity that reflects a change in the public's views of the bird, whose aerial remoteness (how often have you looked a swift in the face, in the same way that you have a sparrow?) and eerie screaming call used to give them rather sinister connotations. Swifts had vernacular names to match, such as Devil Bird (but for a bird with a truly diabolical reputation, read the Nightjar essay).

Even nowadays, their spiky, sword-shaped appearance and shrill noises give them a slight air of the macabre. But we're still sad to see them leave at the end of the summer – our sense of loss sharpened by a primitive, subconscious dread at the impending winter's cold infertility which we associate with their departure. 'Where the Devil are they?' we wonder in the first days of the following spring – and the prehistoric part of us is relieved at their return: the final confirmation that the fecund summer is on its way.

EDIBLE-NEST SWIFTLET

The hundred-dollar bird

Bird's nest soup is one of those dishes which, to people brought up within Britain's notoriously conservative culinary tradition, sounds positively disgusting.

Most of us outside the cognoscenti have an image of a bunch of twigs mixed into a broth. The truth is actually more stomach-churning even than that.

The Edible-Nest Swiftlet, an excellent flier with crescent wings that is slightly smaller than Britain's Common Swift, builds its nests entirely out of its own dried saliva. This is harvested, dissolved in water, and then made into a soup that is a popular delicacy in China. If Britons lie at the extreme conservative end of culinary culture, the Chinese are waving at us rather distantly from the other end.

Gathering the nests of swiftlets that use such tasty saliva, particularly the Edible-Nest Swiftlet, is big business. Hong Kong is the centre of the trade, with the nests then passed on to mainland China or the United States. Good-quality nests can be sold by producers for more than $1,000 a kilo (about eighty nests), and portions of the soup are sold in restaurants for up to $100.

Like most big business, the trade in edible nests contains the seeds of its own destruction. In general, national economies are prone to boom and bust, rather than steady growth, because banks and other companies get greedy and over-extend themselves. The nest industry, which is concentrated in south-east Asia where the birds breed, is no different. For Sarawak's huge Niah Cave colony of Black-Nest Swiftlets in Malaysian Borneo, excessive harvesting has reduced bird numbers from 4.5 million to a mere few hundred thousand birds according to most estimates, though surveying the population is difficult. If the nest goes, the eggs in it are invariably destroyed, and so the colonies decline. It is thought that in Indonesia about 16 million nests are taken every year.

The huge amount of money involved – tens of millions of dollars globally – has inevitably given rise to allegations of shady practices. This ranges from tax evasion all the way to the murder of local people who have tried to harvest the nesting sites, the exploitation of which is often sold by governments to corporations.

The swiftlets have become so lucrative that factory-farming of the birds has begun – with people creating sites designed to attract the birds. This is a controversial practice, but could be what saves them, if governments were to ban the harvesting of nests in the birds' original colonies, and decree that it all had to be done through these artificial locations.

So after all the effort made in attaining it, what does bird's nest soup actually taste like? It is definitely an acquired taste, with a rubbery texture. Debate rages over whether it actually tastes good at all, or is so *recherché* because it is considered an aphrodisiac.

It would be a great pity if the Edible-Nest Swiftlet and its sister species were wiped out. They are fascinating creatures – more like bats than birds in some respects. Those that commonly nest in huge colonies in caves use echoes to work out where they are going and how to get out of the cave. They look rather bat-like too – dull brown creatures with a flight that seems random but is anything but, since they are simply following insects. Scientists have speculated that swiftlets may also use echoes to feed, as well as to do the bird equivalent of finding the keys and leaving the house. If so, this would make them even more like these sky-bound mammals. It may be that swiftlets have taken advantage of the relative absence where they live of the free-tailed bat family – the most aerially acrobatic of all bats – to exploit an underused ecological niche. Birds take opportunities wherever they arise, and those don't only have to be opportunities in the bird world.

There is another reason to fight against the extinction of swiftlets. Their nests are easily harvested because they are colonial breeders – so the caves where they live are like avian factories, where a huge number of nests can be collected very quickly. Why are they colonial? Because they follow concentrated swarms of insects, rather than spreading out to look for them. If your prey hangs around together, you will hang around together. If you specialise in attacking swarms, you are of course one of nature's pest-controllers – a green and healthy equivalent to pesticide. So swiftlets have another, more industrial, use which should benefit humans, aside from the eating of their nests. Nature is a large and complex machine with many working parts. Take away one of them, and the whole thing is liable to stop operating – just listen to the testimony of any bored office worker who has dismantled a ballpoint pen, only to lose a tiny crucial part under the desk before having put it back together again.

WHITE STORK

The bringer of babies

I shall end these essays in topsy-turvy style, by looking at the creature that, according to legend, is the bird of beginnings: the White Stork, which delivers babies to joyful families on its long pointed beak. Why, of all creatures, is this bird credited with such a special power? There are several possible reasons, and I suspect they have all done their bit to buttress this legend.

The first – and turn the page over now if you blush easily – is its phallic appearance. This black-and-white bird has a long, thick, dagger-shaped beak, which looks all the larger because it is bright orange.

Another reason is that if we accept one half of the myth – the notion that babies are delivered from somewhere far away – then it is natural to accept the other half of it: that the White Stork is the natural bird to get the job done. Although scholars debated for centuries whether many species migrated, they (almost) universally accepted that storks did so. This wasn't really a matter for argument because these large birds with huge bright bills are so visible – even the ancients could, without optical aids, see them flying off over the sea. The Old Testament's Book of Jeremiah, written in about the seventh century BC, already noted that 'the stork in the heavens knows its appointed times', in a clear reference to its seasonal movement. Who better to deliver a baby than a bird that has disappeared for several months to an unknown location, only to return in the spring when new life begins?

A further, rather imaginative but perhaps excessively ingenious theory of why the White Stork in particular has been singled out for this mythical task is based on the bird's habits in northern Germany, where the legend is thought to have originated. Modern naturalists point out that the White Stork arrives there about nine months after the Midsummer festival in late June – an occasion when, in pre-Christian times, ancient fertility rites were celebrated. The time lag after Midsummer – roughly the same as a human pregnancy – might have set people thinking.

Language offers another tantalising clue to why the stork is associated with babies. The German for stork is *Storch*. Many scholars think this originally comes from the German for stick – which is *Stock* – because of the stork's predilection for standing on one, thick, stick-like leg, and its rather still, erect posture. 'Stick' or its equivalent also has the slang meaning in many languages, including German, of... well, you can get the picture.

It is ironic that, after centuries of being associated with new life, the stork has been revealed by DNA analysis to be associated with new death. Scientists used to think that storks were closely related to herons, but now say that the New World vultures – those scavengers of carrion – are among their closest relatives. This may seem preposterous, but one very visible clue lies in the fact that some storks (though not the White) have bare skin on the face, in common with many vultures. This is useful if, like a vulture, you spend your life literally stuffing your face into large bloody carcasses and don't want the bother of cleaning your feathers afterwards.

For the White Stork, it doesn't matter how it acquired its reputation for bringing babies, but it is a jolly handy thing to have. Most birds have, over the centuries, been deliberately persecuted at worst, or had their needs carelessly ignored at best. But the persistence of the stork myth has made people eager to protect the bird and even to encourage it to nest near their houses. As a result, although White Storks do not breed in Britain, in parts of Continental Europe the sight of this huge bird sitting on its even bigger nest is a common one even in the very middle of small towns. It is always better to believe in legends, if they do some good.

AFTERWORD

I hope you have enjoyed this book and now regard birds as a little wilier, somewhat more tenacious and a lot more fun than when you turned the first page. The next stage is to help preserve them, because they're a lot more fun alive, singing and bobbing in and out of bushes, than they are as stuffed specimens in glass cases.

The good news is that you can help birds immediately. At over a million members, the Royal Society for the Protection of Birds has more followers than all of Britain's political parties combined – and you could help increase that rather embarrassing gulf by becoming a member.

If you're a hands-on person, there's a lot else you can do besides. You can put out food for the birds in your garden come wintertime – though it's best not to do this in the summer, because it might not be good for their young. You can buy bird tables, baths and nest boxes – and if there's a particular species you want to attract, such as a swift, House Martin or swallow, there's a box designed especially for them. As you grow more experienced, you can join the army of amateurs taking part in official bird surveys organised by the British Trust for Ornithology and other conservation organisations. These surveys are crucial: they reveal which birds are declining, so the professionals can start working out why and what to do about it. I hope this book has shown not just how important birds have been in human life, but vice versa too – the huge power that we wield, often unwittingly, can decide their fates. Over the course of human history we have often persecuted our twittering friends, but we can also be a force for good in the world of birds. Ignoring this would be downright cuckoo.

ACKNOWLEDGEMENTS

My greatest thanks must go to Andy Hayward, a marvel of the publishing world who has always shown much-appreciated faith in me.

Many thanks also to Abbie Headon and Jennifer Barclay at Summersdale – to Abbie for the tough job of editing my book and telling me when my jokes weren't funny, and to Jennifer for her imagination and patience in devising the book's structure.

I am also extremely grateful to Ray Hamilton for his deft and patient copy-editing.

I owe profuse thanks also to my twin brother Christopher, for using his erudite knowledge of birds to conduct a forensic assessment of my manuscript.

My gratitude to Robert Shrimsley, who provided a nook for me at the *Financial Times* where I learnt to write quirky and amusing things and get paid for it.

I am grateful also to Tom Gullick for giving me the most interesting interview in my many years of journalism – if only government ministers were as frank and full of information.

Last but not least, my thanks to that lone singing skylark that inspired my interest in birds back in the 1970s – you are gone but not forgotten.

ABOUT THE AUTHOR

David Turner is a London-based writer who has worked for the *Financial Times* and Reuters. David has been watching birds since the tender age of eight – both in Britain and while a foreign correspondent – and watching people since he graduated with a History degree from Cambridge and became a journalist. He is a volunteer observer for the British Trust for Ornithology.

Alphabetical list of species entries

Have you enjoyed this book?

If so, why not write a review on your favourite website?
Thanks very much for buying this Summersdale book.

www.summersdale.com